RESTORATION VISION for the Laguna de Santa Rosa

APRIL 2020

SFEI San Francisco Estuary Institute

PROJECT DIRECTION
Scott Dusterhoff
Robin Grossinger

AUTHORS
Amy Richey
Scott Dusterhoff
Sean Baumgarten
Emily Clark
Matthew Benjamin
Samuel Shaw

DESIGN AND LAYOUT
Amy Richey
Ruth Askevold
Katie McKnight

PREPARED BY San Francisco Estuary Institute-Aquatic Science Center

IN COOPERATION WITH and FUNDED BY
California Department of Fish and Wildlife
Sonoma Water

IN PARTNERSHIP WITH
Sonoma Water
Laguna de Santa Rosa Foundation

SAN FRANCISCO ESTUARY INSTITUTE PUBLICATION #983

SUGGESTED CITATION

San Francisco Estuary Institute-Aquatic Science Center. 2020. *Restoration Vision for the Laguna de Santa Rosa.* An SFEI-ASC Resilient Landscapes Program report developed in cooperation with Sonoma Water, the Laguna de Santa Rosa Foundation, a Technical Advisory Committee, and a Management Advisory Committee. Publication # 983. San Francisco Estuary Institute-Aquatic Science Center, Richmond, CA.

Version 1.4 (November 2021)

REPORT AVAILABILITY

Report is available at sfei.org

IMAGE PERMISSION

COVER CREDITS

Front cover, left to right: photo, SFEI; imagery courtesy of Google Earth.

Back cover, top to bottom: Field visit in the Laguna with TAC members, photo by SFEI; conceptual cross section of Laguna riparian management in urban areas, photo by SFEI; section from a 1840 "diseño," or sketch, of Rancho Llano de Santa Rosa, showing a string of perennial lakes and ponds along the course of the Laguna de Santa Rosa (Land Case Map B-128, courtesy of The Bancroft Library); and Irwin Creek, photo by SFEI.

THE FEDERATED INDIANS OF GRATON RANCHERIA

The Laguna de Santa Rosa watershed lies within the ancestral territory of the Coast Miwok and the Southern Pomo peoples, whose current and historical cultural legacies are tied inextricably with the native ecosystems and resources described in this document.

Today, the Federated Indians of Graton Rancheria are a congressionally-recognized American Indian tribe. The Graton Rancheria reservation is a sovereign territory and is labeled on the maps in this document as such. While this document shows historical and modern habitats types within the reservation, readers should note that these are from publicly available data sources and may not accurately depict land cover within the territory of the Federated Indians of Graton Rancheria. The reader should contact the Federated Indians of Graton Rancheria to get more detailed information about their land.

Acknowledgments

This project was co-funded by the California Department of Fish and Wildlife (CDFW) Proposition 1 Grant Program, and Sonoma Water. We would like to thank our project partners Neil Lassettre and Mike Thompson of Sonoma Water, Wendy Trowbridge of the Laguna de Santa Rosa Foundation, and Eric Larson of CDFW for lending their support, guidance, and expertise throughout the development of the Vision. We would also like to thank David Kuszmar of the North Coast Regional Water Quality Control Board (NCRWQCB) for helping envision this project.

Members of our Technical Advisory Committee (TAC) provided invaluable technical assistance and guidance through a series of meetings, as well as by providing helpful comments on draft materials. The TAC members were Betty Andrews (ESA), Peter Baye (Coastal and Wetlands Ecologist), Brian Cluer (NOAA), Matthew Cover (CSU Stanislaus), Brenda Grewell (USDA and UC Davis), Eric Larson (CDFW), and Mariska Obedzinski (UCSD, California Sea Grant).

We are very grateful to the Management Advisory Committee (MAC) who participated in meetings and provided helpful comments. The MAC members were Derek Acomb (CDFW), Gina Benigno (CDFW), Lisa Bernard (NCRWQCB), Hattie Brown (Sonoma County Regional Parks), Denise Cadman (City of Santa Rosa), Sierra Cantor (Gold Ridge RCD), Kelsey Cody (NCRWQCB), Clayton Creager (NCRWQCB), Ann DuBay (Sonoma Water), Sheri Emerson (Sonoma County Agricultural Preservation and Open Space District), Noelle Johnson (Gold Ridge RCD), John Mack, (Sonoma County), Alydda Mangelsdorf (NCRWQCB), Greg Martinelli (CDFW), Buffy McQuillen (Federated Indians of Graton Rancheria), Valerie Minton Quinto (Sonoma County RCD), Molly Oshun (Sonoma Water), Jessica Pollitz (Sonoma County RCD), Charles Striplen (NCRWQCB), Liane Ware (City of Rohnert Park), and Chris Watt (NCRWQCB). Additionally, we would like to thank several local landowners who shared their expertise and insights, and provided access to their properties for site visits.

We are grateful to all of the archivists who assisted with historical data collection (see page 43). Thanks to Jenny Blaker for generously sharing materials and assisting with historical data collection and field visits.

Numerous SFEI staff assisted with data collection, compilation, mapping, analysis, reporting, advice, editing, and project management, including Julie Beagle, Gloria Desanker, Steve Hagerty, Meredith Lofthouse, April Robinson, Micha Salomon, Lawrence Sim, Lauren Stoneburner, and Patrick Walsh.

Contents

Acknowledgments iii

Executive Summary ix

1 Introduction 1

Background 1

Laguna Restoration Vision 11

2 Laguna Restoration Vision Process 17

The Process for Developing a Vision 17

3 Overview of Changes in Key Drivers of Ecosystem Structure & Function 23

Introduction 23

Changes in Hydrology 23

Changes in Sediment Dynamics 30

Changes in Nutrient Dynamics 34

4 Historical Ecology of the Laguna de Santa Rosa 41

Introduction 41

Methods 42

Results 46

Summary 61

5 Landscape Change Analysis 63

Introduction 63

Methods 65

Changes in Habitat Extent 68

Changes in Habitat Configuration 74

Introduction of Invasive Species 89

Summary 93

6 Future Conditions and Climate Change Predictions 95

Rising Temperatures and Extreme Heat Events 95

Precipitation Patterns 96

Shifting Vegetation Patterns 96

Drought 97

Wildfire 98

Building Climate Resilience 98

7 Restoration Vision 101

Laguna de Santa Rosa Restoration Vision Map 103

Vision Concepts 104

Benefits of Implementing the Landscape Vision 118

Watershed-Scale Management for Supporting the Vision 130

8 Key Considerations and Known Challenges 143

9 Synthesis and Next Steps 145

Synthesis 145

Next Steps 148

10 References 150

Appendix A 168

List of Additional California Native Plants Designated as Species of Local Concern

Appendix B 169

List of California Native Wetland Plants Found in the Historical Record

Appendix C 176

Detailed Methods

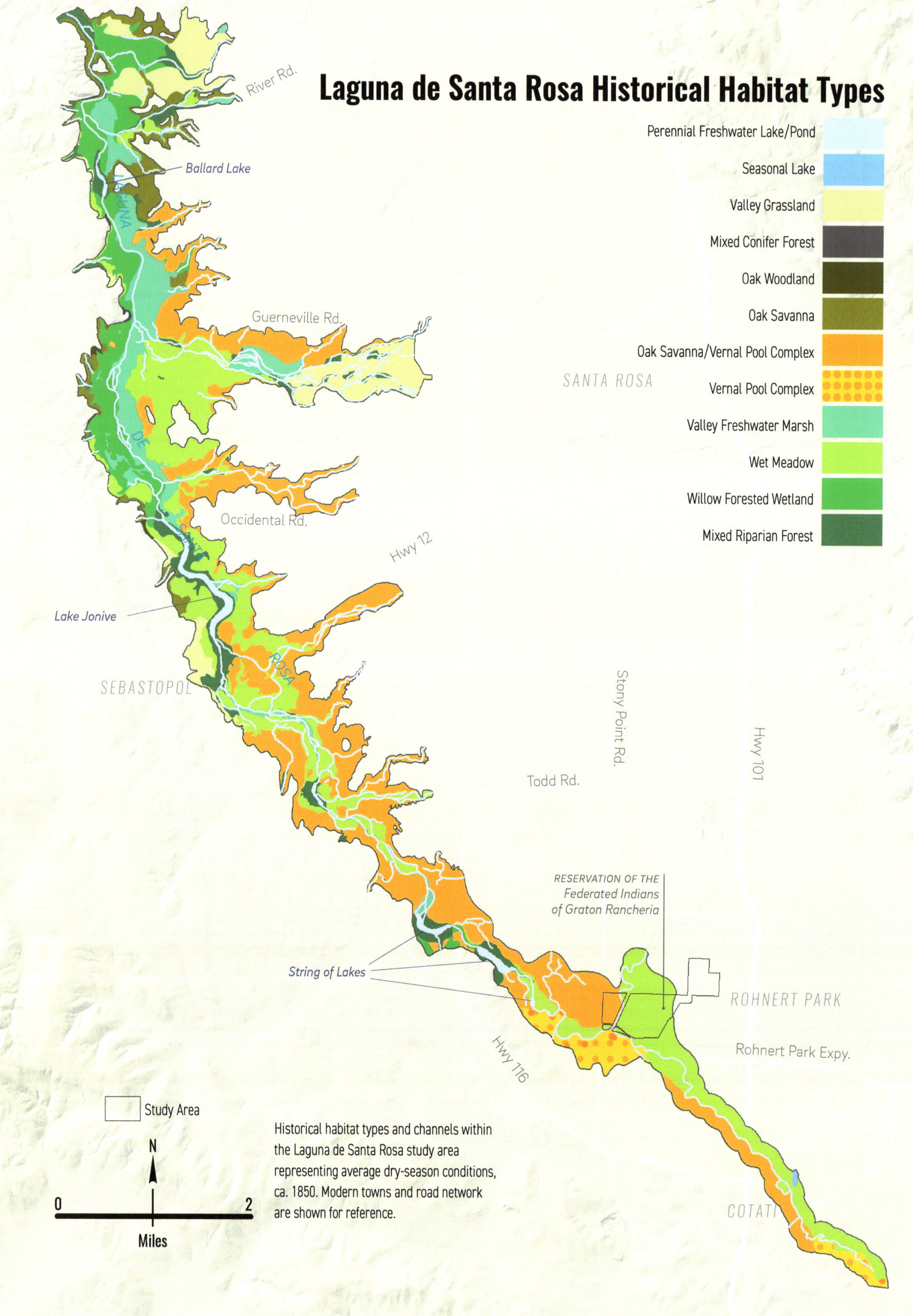

Historical habitat types and channels within the Laguna de Santa Rosa study area representing average dry-season conditions, ca. 1850. Modern towns and road network are shown for reference.

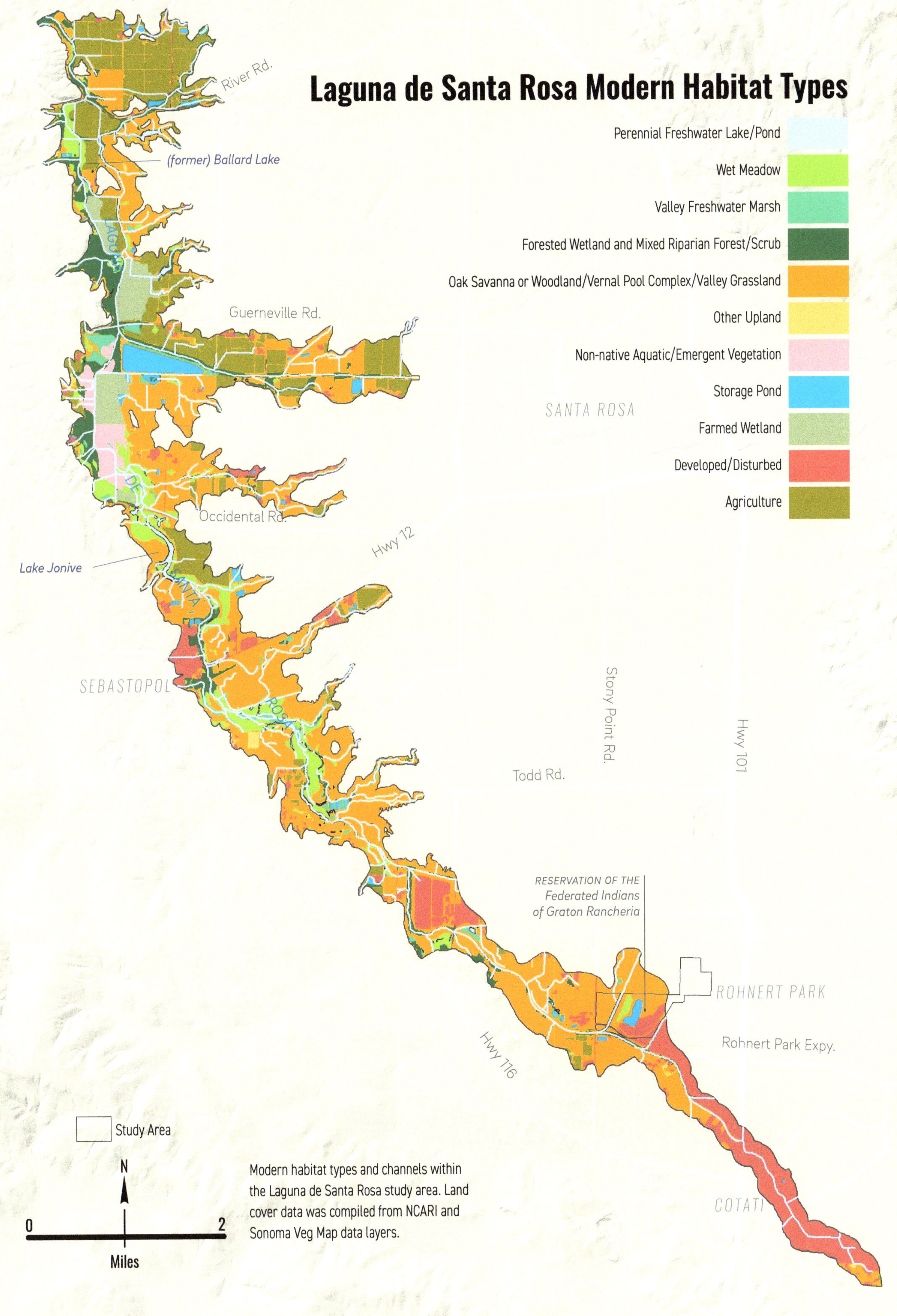

Modern habitat types and channels within the Laguna de Santa Rosa study area. Land cover data was compiled from NCARI and Sonoma Veg Map data layers.

Executive Summary

The Laguna de Santa Rosa, located in the Russian River watershed in Sonoma County, CA, is an expansive freshwater wetland complex that hosts a rich diversity of plant and wildlife species, many of which are federally or state listed as threatened, endangered, or species of special concern. The Laguna is also home to a thriving agricultural community that depends on the land for its livelihood. Since the mid-19th century, development within the Laguna and its surrounding watershed have had a considerable impact on the landscape, affecting both wildlife and people. Compared to pre-development conditions, the Laguna currently experiences increased stormwater runoff and flooding, increased delivery and accumulation of fine sediment and nutrients, spread of problematic invasive species, and decreased habitat for native fish and wildlife species. Predicted changes in future precipitation patterns and summertime air temperatures, combined with expanding development pressure, could exacerbate these problems. People who manage land and regulate land management decisions in and around the Laguna, including landowners; federal, state, and local agencies; and local stakeholders, are seeking a long-term management approach for the Laguna that improves conditions for the wildlife and people that call the Laguna home. The California Department of Fish and Wildlife and Sonoma Water funded the Laguna-Mark West Creek Watershed Master Restoration Planning Project to develop such a management approach, focusing on the need to identify restoration and management actions that enhance desired ecological functions of the Laguna, while also supporting the area's agriculture and its local residents.

View of the Laguna de Santa Rosa. Imagery: Google Earth.

The first step in the Restoration Planning Project was the development of a long-term Resilient Landscape Vision within the Laguna's 100-year floodplain that highlights opportunities for multi-benefit habitat restoration and land management. Vision development began by establishing an understanding of the landscape function from past, present, and potential future perspectives. This included high-level syntheses of existing published and unpublished information regarding conditions in the Laguna's surrounding watershed, and original detailed technical analyses focused on constructing a picture of the historical ecology of the Laguna, as well as the magnitude of change in habitat conditions over the past two centuries. Key findings from the technical analyses include the following:

Historical Ecology of the Laguna - Prior to major European American modification, the Laguna was characterized by a diverse and extensive array of wetland, riparian, and aquatic habitats that supported a wide variety of plants and animals, and provided an abundance of resources for native peoples. A series of deep, cold lakes marked the path of the Laguna along the Sebastopol Fault, and provided habitat for salmonids and other native fish and wildlife. Mixed riparian forests dominated by oaks, willows, Oregon ash, and other species surrounded many of these lakes, as well as portions of the Laguna mainstem and tributary channels. Perennial wetlands, including valley freshwater marsh and willow forested wetland, formed large expanses downstream of present-day Occidental Road where high groundwater maintained perennially saturated soils. Slightly drier areas on the surrounding floodplain supported seasonal wetlands such as wet meadows and vernal pool complexes. The highly productive wetlands of the Laguna supported a complex food web including a diversity of plants and huge numbers of resident and migratory birds, fish, reptiles, amphibians, and mammals.

Landscape Change - Though the Laguna still provides valuable wildlife habitat and range of other ecosystem services, the landscape has been heavily modified over the past two centuries. Early hunting pressures resulted in the extirpation of a number of native wildlife species from the watershed, while grazing and cultivation resulted in the conversion of large portions of the Santa Rosa Plain to agricultural uses. Many portions of the Laguna were drained and filled for agriculture, and many of the tributaries feeding the Laguna (as well as portions of the mainstem) were channelized and/or re-routed for flood control. Urban and agricultural development on the Santa Rosa Plain have led to further habitat loss and major changes in hydrology and sediment dynamics, while increased nutrient inputs to the Laguna have impaired water quality and contributed to the expansion of *Ludwigia hexapetala* and other invasive species. These land and water use modifications have resulted in a ~60% decrease in wetland habitat and widespread habitat fragmentation, reducing the Laguna's ability to support native biodiversity and provide other ecological functions.

Between March 2018 and May 2019, the project team held a series of meetings to present information on the past, present, and potential future landscape functioning and begin development of the Resilient Landscape Vision for the Laguna. The meetings were

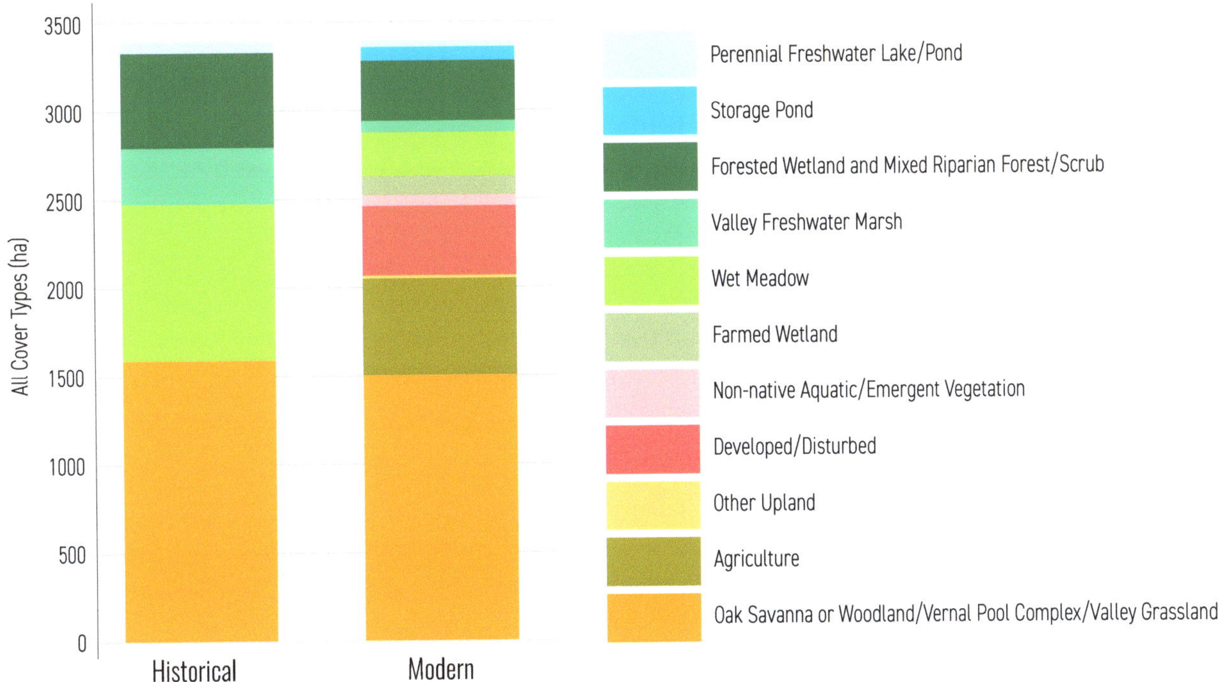

Change in extent of each land cover type within the study area between historical (ca. 1850; bar on left) and modern (ca. 2015; bar on right) time periods.

attended by a team of technical advisors and a group of stakeholders that included state regulatory agencies, county permitting agencies, county land management agencies, local municipalities, local land management agencies, local nonprofit organizations, and local residents. Based on the information presented at the meetings and the feedback provided by the technical advisors and stakeholders, the project team identified a suite of short-term and long-term restoration and management concepts that together form the Resilient Landscape Vision. The Vision concepts include the following:

Wetland and Aquatic Habitat Restoration - Restoring open water, freshwater marsh, wet meadow, willow forested wetland, mixed riparian forest, and oak savanna/vernal pool habitats throughout the Laguna;

Riparian Management - Levee setback and channel realignment along lower Mark West Creek and lower Santa Rosa Creek, vegetation enhancement, invasive species management, and native planting in tributary channels and in the urbanized portion of the upper Laguna;

Infrastructure Redesign - Redesign of bridges with larger spans to convey greater flows and allow more room for wildlife, prioritizing the Guerneville and Occidental road bridges; long-term redesign of wastewater treatment infrastructure to move it further from the Laguna mainstem.

Implementing the Vision would transform the landscape and bring back lost habitat features and ecosystem functions, which would support a variety of services for the wildlife and people that live within and around the Laguna. These functions include life history support for wildlife (e.g., birds, fish, amphibians, and reptiles), nutrient and pollution regulation, water temperature regulation, flood management, and increased recreational and aesthetic value. Specific outcomes associated with Vision implementation include the following:

Significant increases in wetland land cover types - Compared to the current landscape, the Vision would lead to an approximate 20% increase in open water habitat, an approximate 50% increase in mixed riparian forest habitat, and a doubling of valley freshwater marsh and wet meadow habitats within the Laguna;

Larger, more connected freshwater marshes and wet meadows - The Vision would lead to an overall increase in the extent of large patches (i.e., discrete areas between 10 and 500 hectares) of freshwater marsh and wet meadow habitats, and increased connectivity (i.e., decreased distance) between large patches and smaller patches;

HISTORICAL, MODERN, and VISION WETLAND COVER

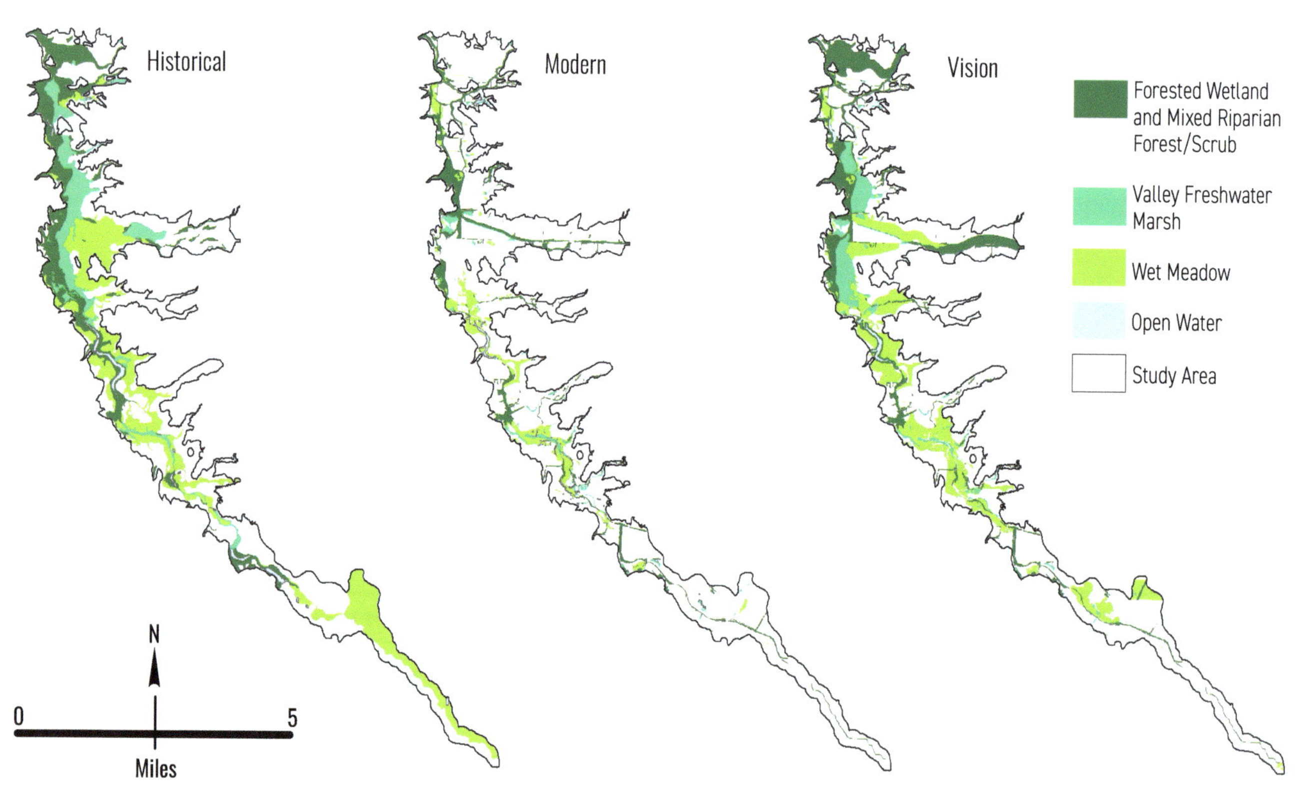

Increased riparian buffer width and connectivity - Vision implementation would increase the total channel length with riparian forest by approximately 10%, the proportion of the channel network with wide riparian areas (i.e., areas with a riparian width greater than 100 m) would increase threefold (10% to 30% of the total channel length), and the overall extent of large riparian patches and their connectivity with smaller patches would increase;

Greater extent of wetlands adjacent to channels - Implementation of the Vision would lead to an overall increase in the proportion of Laguna channels adjacent to wetland habitats, with the total channel length adjacent to valley freshwater marsh and wet meadow habitats increasing by two-thirds;

Greater proportion of natural land cover types in terrestrial zones around wetlands - Under the Vision, the proportion of wetlands in the Laguna surrounded by natural land cover types would increase from approximately 50% to 60%, mostly due to a five-fold increase in the area of riparian vegetation adjacent to wetlands.

The next step in the Laguna-Mark West Creek Watershed Master Restoration Planning Project is the development of a Restoration Plan that builds from the Restoration Vision. The Restoration Plan will describe near-term restoration targets and project concepts developed from the Vision that help meet the targets. The targets and concepts will be determined in close coordination with the project technical advisors, stakeholders, and local landowners. The Restoration Plan will provide details about the process for developing the project concepts, the habitat features within each project concept, each project concept's contribution toward meeting the near-term restoration targets, and the recommended order of implementation. The ultimate success of the Vision and the resulting Restoration Plan will be driven by local landowner support and adequate funding for implementation, as well as an updated approach to landscape management within the Laguna's surrounding watershed. Specifically, the long-term viability of habitat restoration in the Laguna will largely depend on changes to the delivery of flow (surface and subsurface), fine sediment, and nutrients to the Laguna. §

Education in the Laguna. Photo: Laguna de Santa Rosa Foundation.

Laguna de Santa Rosa in early spring. Photo: SFEI.

1 Introduction

Background

The Laguna de Santa Rosa is an expansive freshwater wetland complex in the Russian River watershed that is the most biologically diverse region of Sonoma County, and the second largest freshwater wetland complex in Northern California (PWA 2004a, Laguna de Santa Rosa Foundation 2011). The Laguna includes seasonal and perennial creeks, ponds, wet meadows, marshes, vernal pools, forested riparian areas, oak woodlands, and grasslands. Its complex habitats are home to over 200 species of birds ranging from song sparrows *(Melospiza melodia)* to common yellowthroat *(Geothlypis trichas)* to bald eagles *(Haliaeetus leucocephalus)*, and it is a major stopover for thousands of waterfowl as they traverse the Pacific Flyway. The Laguna supports many species of mammals, from river otters *(Lontra canadensis)* to bobcats *(Lynx rufus)* to mountain lions *(Puma concolor)*, as well as reptiles and amphibians such as western pond turtle *(Actinemys marmorata)*, California tiger salamander (*Ambystoma californiense*), and gopher and garter snakes (*Pituophis catenifer, Thamnophis sirtalis*). In addition to many native and non-native warm water fishes that live in the Laguna year-round, its waterways also support migrating coho salmon (*Oncorhynchus kisutch*) and steelhead (*Oncorhynchus mykiss*), and provide vital feeding and rearing habitat for these salmonids (Honton and Sears 2006, USFWS 2016). Its plant diversity reflects a high degree of richness, hosting both common species, such as valley oaks (*Quercus lobata*) and willows (*Salix* spp.), as well as unique endemic species in its vernal pools and wetlands (USFWS 2005). Because of its importance for a wide array of terrestrial and aquatic wildlife, a portion of the Laguna is recognized as a Ramsar Wetland of International Significance (Sloop 2009).

Although it remains a vital ecosystem, the Laguna has been considerably altered since the onset of intensive European-American settlement in the region and has the potential to experience more issues in the future. Over the past 200 years, development within the Laguna and its surrounding watershed have had a considerable impact on the landscape, affecting conditions for both wildlife and people. Urbanization, conversion of lands to agriculture, and the rerouting and channelization of rivers and creeks that drain to the Laguna have led to problems such as increased stormwater runoff and flooding in the Laguna, increased fine sediment and nutrient delivery to and accumulation within the Laguna, introduction of invasive species, and widespread habitat loss for native fish and wildlife species within and adjacent to the Laguna. Looking to the future, predicted changes in precipitation patterns and summertime air temperatures, combined with expanding development pressure, will likely exacerbate many of these problems. Addressing the issues in the Laguna and its surrounding landscape that affect both wildlife and people will require a management approach that considers how the ecosystem has changed since the onset of intensive development and the changes that could be coming as the population in the area continues to grow and climatic conditions continue to shift.

Over the past few decades, there have been many efforts focused on addressing the challenges the Laguna faces, and the appropriate management approaches for preserving and enhancing the ecosystem. Foundational work includes *Enhancing and Caring for the Laguna de Santa Rosa* (Honton and Sears 2006), which identifies key focus areas for improving the Laguna ecosystem, and *The Altered Laguna: A Conceptual Model for Watershed Stewardship* (Sloop et al. 2007), which lays out a conceptual understanding of the past and present physical and biological functioning of the Laguna and management recommendations for ecosystem improvement. Additionally, the US Environmental Protection Agency funded sediment and nutrient studies in support of Total Maximum Daily Load (TMDL) development that provide detailed information on past and present sediment and nutrient dynamics in the Laguna (Tetra Tech 2015a; 2015b). More recently, historical ecology studies have investigated historical alignments of tributaries to the Laguna, estimated historical and modern nutrient loads, and investigated land cover in the central portion of the Laguna (Butkus 2010, 2011b; Dawson and Sloop 2010; Baumgarten et al. 2014, 2017). Additionally, a partnership of state and local entities identified management actions within the Laguna that could improve overall ecosystem functioning (SCWA et al. 2016). Although these

(Left) Song sparrow (*Melospiza melodia*), Photo: Becky Matsubara; California tiger salamander (*Ambystoma californiense*), Photo: US Fish and Wildlife Service; bobcat (*Felis rufus*), Photo: Becky Matsubara; and steelhead (*Onchorrynchus mykiss*), Photo: NOAA Fisheries.

List of Impaired Water Bodies in the Laguna Watershed

List of impaired water bodies by hydrologic sub-area (HSA) segment in the Laguna de Santa Rosa and its contributing watershed, and by pollutant (Indicator Bacteria, Dissolved Oxygen, Mercury, Phosphorus, Sedimentation/Siltation, Temperature). From the California State Water Board – see https://www.waterboards.ca.gov/northcoast/water_issues/programs/watershed_info/russian_river/laguna_de_santa_rosa/.

WATER BODY NAME	LISTED SEGMENT	POLLUTANT
Laguna HSA mainstem Laguna de Santa Rosa	Entire Water Body	Indicator Bacteria
		Oxygen, dissolved
		Mercury
		Phosphorus
		Sedimentation/Siltation
		Temperature
Laguna HSA tributaries to the Laguna de Santa Rosa (except Santa Rosa Creek and its tributaries)	Mainstem Colgan Creek	Oxygen, Dissolved
	Entire Water Body	Indicator Bacteria
		Sedimentation/Siltation
		Temperature
Mark West HSA mainstem Mark West Creek downstream of the confluence with the Laguna de Santa Rosa	Entire Water Body	Aluminum
		Oxygen, Dissolved
		Phosphorus
		Manganese
		Sedimentation/Siltation
		Temperature
Mark West HSA mainstem Mark West Creek upstream of the confluence with the Laguna de Santa Rosa	Entire Water Body	Sedimentation/Siltation
		Temperature
Mark West HSA tributaries to Mark West Creek (except Windsor Creek and its tributaries)	Entire Water Body	Sedimentation/Siltation
		Temperature
Mark West HSA Windsor Creek and its tributaries	Entire Water Body	Sedimentation/Siltation
		Temperature
Santa Rosa HSA mainstem Santa Rosa Creek	Entire Water Body	Indicator Bacteria
		Sedimentation/Siltation
		Temperature
Santa Rosa HSA tributaries to Santa Rosa Creek	Spring Lake	Mercury
	Entire Water Body	Indicator Bacteria
		Sedimentation/Siltation
		Temperature

and many other efforts represent a tremendous amount of dedication by many people focused on approaches to managing the Laguna, there is still more to be done. Specifically, there is not yet a cohesive restoration plan that synthesizes the best available knowledge of past, present, and potential future ecosystem conditions and provides specifics about restoration efforts within and adjacent to the Laguna. There is a particular need to consider both current and projected future land use and climatic conditions when developing restoration guidelines that seek to generate ecosystem benefits.

In 2017, Sonoma Water, in partnership with the Laguna de Santa Rosa Foundation and the San Francisco Estuary Institute (SFEI), received a Proposition 1 grant from the California Department of Fish and Wildlife (CDFW) for a project titled *Laguna-Mark West Creek Watershed Master Restoration Planning Project*. The overall goal of this 4-year effort is to develop a plan that supports ecosystem services in the Laguna—through the restoration and enhancement of landscape processes that form and sustain habitats and improve water quality—while considering flood management issues and the productivity of agricultural lands. The project is focused primarily on the Laguna within the current FEMA-defined 100-year floodplain, but also considers management actions in the surrounding watershed that are necessary for supporting restoration within the Laguna (Fig. 1-1, 1-2, following pages). The project is comprised of three main elements:

- **Restoration Vision -** A long-term landscape-scale vision describing opportunities for restoring lost ecosystem functions and vital habitats throughout the Laguna that combined define the long-term overall habitat restoration goal
- **Restoration Plan -** A detailed plan built from the Restoration Vision that identifies near-term habitat restoration goals and describes a suite of site-scale restoration projects in the Laguna developed from Vision that help meet the near-term goals
- **Restoration Project Designs -** Preliminary designs for a set of high priority projects presented in the Restoration Plan that can be developed along with appropriate California Environmental Quality Act (CEQA) compliance documentation and permit applications

Once completed, the project will provide project partners and stakeholders with restoration project designs that can be moved forward toward implementation, and will provide all entities that own and manage land within and around the Laguna with a clear roadmap for the restoration actions needed to improve the Laguna ecosystem. Because much of the land within and around the Laguna is privately owned, the ultimate success of restoration planning and implementation in and around the Laguna will depend in large part on the will and support of local residents.

Abbreviations

FE = Federally Endangered
FT = Federally Threatened
SE = California Endangered
ST = California Threatened
SC = Federal species of special concern
CSC = California species of special concern

CNPS RANKINGS
1B.1 = Seriously threatened in California
1B.2 = Moderately threatened in California

* = included in Santa Rosa Plain Conservation Strategy

Federal & State Listed Species

Dozens of species, both common and rare, compose the biodiversity of the Laguna. Some species merit special attention to ensure thier continued persistence. Below is a list of species present in the Laguna study area and surrounding watershed that have been identified for protection under Federal or State laws and guidelines. The California Native Plant Society (CNPS) has designatied additional plant species of local concern (Appendix A).

	Common Name	Species Name	Federal Status	State Status	CNPS Rank
Invertebrates	California Freshwater Shrimp	*Syncaris pacifica*	FE	SE	–
Fish	Steelhead Trout	*Onchorhyncus mykiss*	FT	–	–
	Coho Salmon	*Onchorhyncus kisutch*	FE	–	–
	Chinook Salmon	*Onchoryncus tsawytscha*	FT	–	–
	Hardhead	*Mylopharodon conocephalus*	–	CSC	–
	Russian River Tule Perch	*Hysterocarpus traskii* ssp. *pomo*	–	CSC	–
Amphibians and Reptiles	California Tiger Salamander*	*Ambystoma californiense*	FT	CSC	–
	Red-legged Frog	*Rana aurora draytonii*	FT	CSC	–
	Western Pond Turtle	*Emys murmurata*	–	CSC	–
Birds	Bald Eagle	*Haliacetus leucocephalus*	FT	SE	–
	Western Yellow-billed Cuckoo	*Cocczyus americanus occidentalis*	–	SE	–
	Northern Spotted Owl	*Strix occidentalis caurina*	FT	–	–
	Willow Flycatcher	*Empidonax traillii*	–	SE	–
Mammals	Long-eared Myotis	*Myotis evotis*	SC	–	–
	Fringed Myotis	*Myotis thysanodes*	SC	–	–
	Yuma Myotis	*Myotis yumanensis*	SC	–	–
	Townsend's Big-eared Bat	*Corynorhinus tonsendii*	–	CSC	–
	Pallid Bat	*Antrozous pallidus*	–	CSC	–
	Ringtail	*Bassaricus astutus*	–	CSC	–
Plants	Sonoma Alopecurus	*Alopecurus aequalis* var. *sonomensis*	FE	–	1B.1
	Vine Hill Manzanita	*Arctostaphylos densiflora*	–	SE	1B.1
	Sonoma Sunshine*	*Blennosperma bakeri*	FE	SE	1B.1
	White Sedge	*Carex albida*	FE	SE	–
	Sonoma Spineflower	*Chorizanthe valida*	FE	SE	1B.1
	Vine Hill Clarkia	*Clarkia imbricata*	FE	SE	1B.1
	Burke's Goldfields*	*Lasthenia burkei*	FE	SE	1B.1
	Pitkin Marsh Lily	*Lilium pardalinum* ssp. *pitkinense*	FE	SE	1B.1
	Sebastopol Meadowfoam*	*Limnanthes vinculans*	FE	SE	1B.1
	Many-flowered Navarretia*	*Navarretia leucocephala* ssp. *plieantha*	FE	SE	1B.2
	Hickman's Cinquefoil	*Potentilla hickmanii*	FE	SE	1B.1
	Showy Indian Clover	*Trifolium amoenum*	FE	–	1B.1

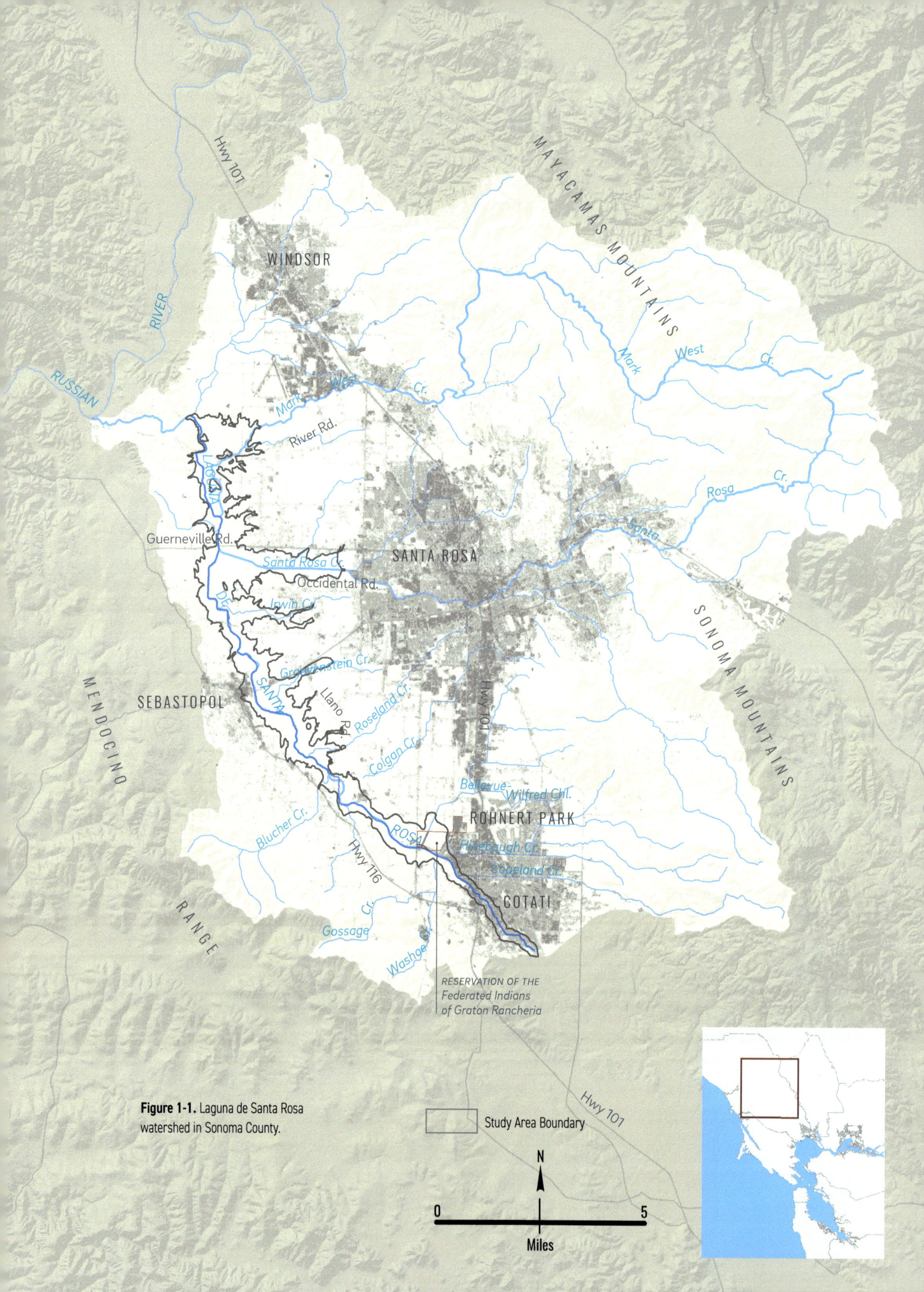

Figure 1-1. Laguna de Santa Rosa watershed in Sonoma County.

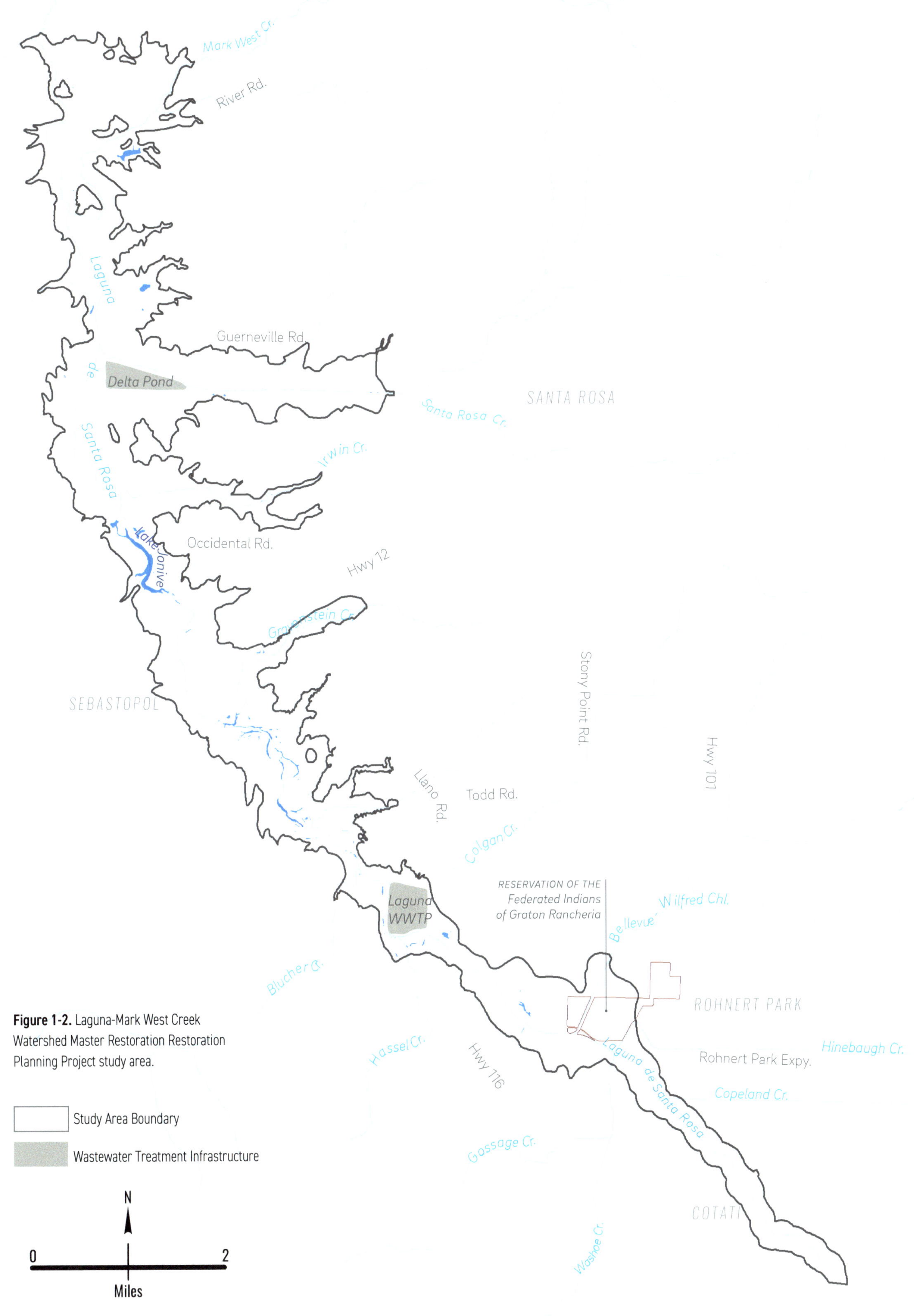

Figure 1-2. Laguna-Mark West Creek Watershed Master Restoration Restoration Planning Project study area.

MANAGEMENT GOALS THE PROJECT ADDRESSES

The following Management Goals are the desired ecosystem outcomes for the Laguna that the Restoration Vision and Restoration Plan aim to address:

Improve overall Laguna ecosystem functions and services for people and wildlife. Biogeochemical, hydrological, geomorphic, and ecological processes support ecosystem functions and services such as generation and preservation of soil, provision of food, recycling of nutrients, filtration of waste and pollutants, climate moderation, flood management, and maintenance of the hydrological cycle that support people and wildlife (Folke et al. 1996, Folke et al. 2004, MEA 2005, de Bello de al. 2010). Improving desired ecosystem functions and services in the Laguna over the long term depends on addressing the following:

- Flooding and groundwater dynamics, including flood frequency and magnitude, and groundwater recharge;
- Sediment dynamics, addressing the current issue of excess sediment delivery to the Laguna;
- Nutrient dynamics, addressing loading and concentration of nutrients in Laguna water and soil;
- Habitat support for desirable wildlife, including native salmonids and other fishes, resident and migratory waterfowl and birds, and terrestrial wildlife, including riparian- and wetland-dependent plant and animal species.

Establish a landscape that will be resilient under a changing climate. Though there is a range of projected futures under a changing climate, it is reasonable to conclude that future climate conditions in Sonoma County will be hotter and drier, and that precipitation and runoff will be more unpredictable and likely of larger magnitudes than historically (see Chapter 6). These changes will profoundly influence ecological outcomes for the Laguna, and there is a desire to ensure that the Laguna landscape can continue to sustain ecological services despite these projected stresses and perturbations.

Enhance environmental, cultural, and agricultural benefits of current and future land uses within and adjacent to the Laguna. In addition to physical and ecological facets of the Laguna system, it is important to consider the importance of supporting the different ways people use the land. Local communities include native people with extensive traditional ecological knowledge of the Laguna and others whose families have lived in and around the Laguna for generations. Many residents look to the Laguna as a place to live and to thrive, including using the land to farm and ranch, to enjoy natural landscapes and wildlife, and to continue cultural practices from one generation to the next. The actions of people in the Laguna watershed, including trash deposition, habitat restoration, sediment and nutrient delivery to the Laguna, and more, can affect every aspect of management. The development and implementation of restoration projects can help rally communities and individuals around the value of restoration, as well as enhance cultural and agricultural benefits.

MANAGEMENT OBJECTIVES THE PROJECT ADDRESSES

The following Management Objectives are conditions that must be attained to accomplish the Management Goals.

- Mimic a natural hydrograph in lands draining to the Laguna that can decrease stormwater velocity and discharge to the Laguna during frequently occurring storm events, and increase groundwater recharge.
- Decrease sediment and nutrient delivery to the Laguna, especially at areas of high deposition/accumulation rates. Move sediment from accumulation areas where appropriate.
- Enlarge riparian and wetland habitat patches and improve their connectedness.
- Control the extent of invasive plant species, and encourage conditions that enable native species to outcompete invasives (e.g., Ludwigia spp., emerging invasive species).
- Improve late spring/summer water quality through improved drainage and flow conveyance.

Young oak planted near the Laguna. Photo: SFEI.

Hooded mergansers. Photo: SFEI.

Sonoma sunshine. Photo: SFEI.

River otter. Photo: USFWS.

Coho salmon. Photo: Oregon Department of Fish and Game.

Western pond turtle. Photo: US Forest Service.

Laguna Restoration Vision

OVERVIEW

This report details the Restoration Vision developed for the Laguna. The Vision is built from new technical analyses and a synthesis of existing information, and was developed in close coordination with a group of technical advisors, agency stakeholders, and landowners. The structure of this report is as follows:

- **Process for developing the Vision (Chapter 2) -** Provides an overview of the process SFEI and project partners followed to compile the necessary information, build the Vision, and determine the ecosystem benefits associated with the opportunities shown in the Vision
- **Landscape change over the past 200 years (Chapters 3-5) -** Includes a high-level overview of changes to key drivers of habitat conditions in the Laguna, a detailed description of historical Laguna habitats (ca. 1850s), and a detailed description of the magnitude of Laguna habitat change from past to present
- **Projected climate change impacts (Chapter 6) -** Provides a general description of the possible ecosystem impacts associated with climate change-driven shifts in precipitation and air temperature
- **Restoration Vision for the Laguna (Chapter 7) -** Provides a map of the restoration opportunities, which were identified through synthesis of the information presented in Chapters 3 through 6 and discussions with technical advisors and stakeholders; detailed information about the restoration opportunity types and associated ecosystem benefits; and a quantification of the increase in habitat extent that could be achieved by implementing the Vision
- **Considerations and constraints associated with Laguna restoration (Chapter 8) -** Discusses known considerations and constraints that need to be addressed to move the Vision restoration opportunities towards design and implementation
- **Next steps (Chapter 9) -** Discusses the next steps for this project and the process of developing the Vision into the Restoration Plan

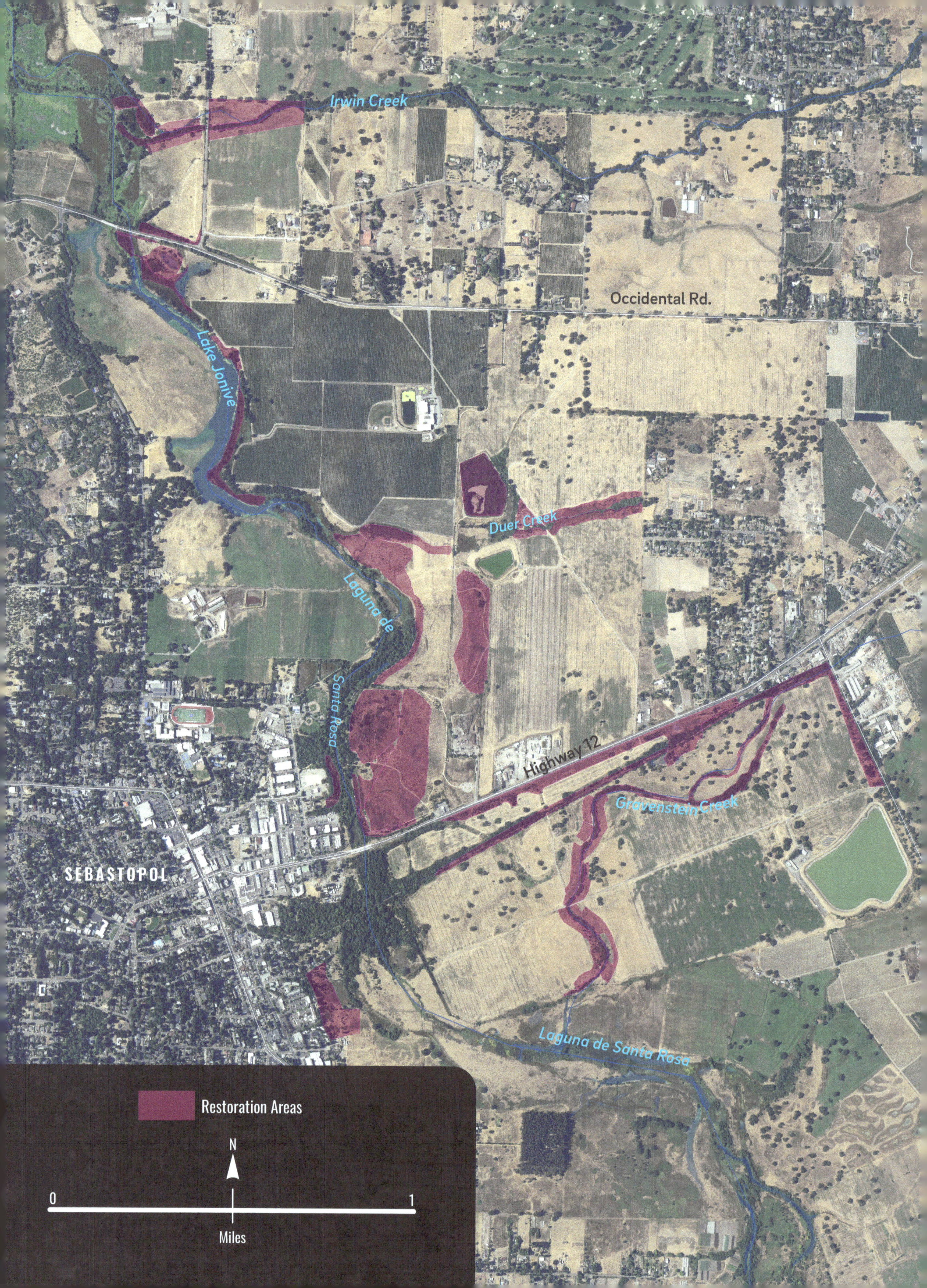
Irwin Creek
Occidental Rd.
Lake Jonive
Duer Creek
Laguna de
Santa Rosa
Highway 12
Gravenstein Creek
SEBASTOPOL
Laguna de Santa Rosa
Restoration Areas
N
0
1
Miles

A VISION ENACTED: Laguna Middle Reach

Over the past decades, multiple partners have come together with the common goal to improve habitat conditions in the middle portion of the Laguna. Extending about 5 miles from the confluence of Gravenstein Creek in the south to Irwin Creek in the north, the Laguna Middle Reach hosts public lands administered by the City of Santa Rosa, the City of Sebastopol, CDFW, Sonoma County Regional Parks, Sonoma County Agricultural Preservation and Open Space District, and the Laguna de Santa Rosa Foundation.

In the late 19th century, this part of the Laguna hosted a resort along Lake Jonive that attracted boaters and swimmers. Habitat types along this reach were a complex mix of open water, mixed riparian forest, freshwater marsh, wet meadow, oak savanna, and willow forested wetland. By the early 20th century, land uses in this area included discharges of raw sewage from Sebastopol into Lake Jonive, as well as agricultural conversion and development of an airstrip that was later used as a dump site for apple production waste. In the mid-20th century, fish kills due to poor water quality were common in Lake Jonive and in the surrounding Laguna. Swimming conditions were very poor, with the presence of raw sewage contributing to a polio outbreak in 1943. In the mid-1940s, efforts to improve water quality and to increase farmland adjacent to the Laguna included dredging to facilitate increased drainage and flushing. Still, by the early 1970s, the area was so heavily impacted and wildlife populations so depleted, that people took further action. Sewage was directed to the Laguna treatment facility, greatly improving water quality, and the City of Santa Rosa purchased four farms, using the land to filter a portion of its recycled water, and to improve habitat.

Together with the help of volunteers, public agencies and nonprofits have teamed to restore more than 180 acres of riparian forest, wet meadow, and oak woodland. Activities have included demonstration wetlands at Kelly Farm, riparian tree planting to widen the riparian buffer along Irwin and Gravenstein Creeks, weed management, and tree planting for shade and habitat along the eastern side of Lake Jonive. Improvements to public trail access have included the addition of seven miles of trails, including the Joe Rodota Trail and the Laguna Discovery trail (Cummings 2004, Laguna de Santa Rosa Foundation 2016).

Volunteers planting at Duer Creek in 2008. Photo: Laguna de Santa Rosa Foundation.

Planted oak trees at Duer Creek in the fall of 2019. Photo: SFEI.

PRIOR EFFORTS THAT HELPED BUILD THE VISION

The Vision greatly benefits from a wide variety of prior research and management efforts. The efforts described here represent those that were relied upon heavily for this effort to develop an understanding of the physical and biological processes at play in the Laguna, to show how those processes have changed over time or are projected to change, and to shape the focus of restoration actions aimed at preserving and enhancing the ecosystem.

Enhancing and Caring for the Laguna

The Laguna de Santa Rosa Foundation produced this foundational work to summarize the physical and ecological factors that influence the Laguna and outline goals for its restoration (Honton and Sears 2006). The report puts forth seven key focus areas in which to improve the long-term health of the Laguna: habitat restoration, ecological research, flood management, stream channel improvements, sedimentation reduction, water quality enhancements, and recreational access and trail development. To identify these focus areas and specific goals within them, the Laguna de Santa Rosa Foundation leveraged input from local ranches and farms, public regulatory and enforcement agencies, regional resource conservation districts, and various other stakeholder groups.

The Altered Laguna: A Conceptual Model for Watershed Stewardship

In support of analyses needed to set TMDLs for Laguna water quality, *The Altered Laguna* presents a series of conceptual models in three categories: hydrology and sedimentation, water quality, and ecology (Sloop et al. 2007). These conceptual models provide a framework for answering key management questions. The report summarizes contemporary information about how the Laguna works, identifies uncertainties and data gaps, and develops recommendations for addressing pressing management needs in the areas of habitat restoration, flood protection, water quality, and water management.

Research Supporting the TMDL

Since the Laguna was first added to California's 303(d) list of impaired water bodies, numerous studies have occurred to inform setting TMDLs for the waterway. Three relatively recent reports provide overviews of this body of research: Fitzgerald et al.'s "Summary of TMDL Development Data Pertaining to Nutrient Impairments in the Laguna de Santa Rosa Watershed" (2013), Tetra Tech's *Laguna de Santa Rosa Nutrient Analysis* (2015a), and Tetra Tech's *Laguna de Santa Rosa Sediment Budget* (2015b). Fitzgerald et al. summarizes several decades of water quality monitoring data for nitrogen, phosphorus, dissolved oxygen, temperature, and sediment in the Laguna from the beginning of each monitoring effort through 2010. Tetra Tech (2015a) outlines the sources of nutrients and organic material in the Laguna and details the modeling efforts conducted to better understand the Laguna's nutrient-related impairments. Tetra Tech (2015b) provides an assessment of past and present sediment delivery to and deposition within the Laguna to better understand the Laguna's fine sediment-related impairment.

United States Geological Survey (USGS) Research

Various publications from researchers at the USGS have illuminated key physical, biogeochemical, and climatic factors influencing the Laguna's hydrology. Nishikawa et al. (2013) documented aspects of the Santa Rosa Plain's surface water and groundwater hydrology, hydrogeology, and water quality, and Woolfenden and Nishikawa (2014) built upon this study to develop a coupled groundwater and surface water model for the region. Curtis et al. (2013) analyzed the spatial distribution of past and present sediment deposition within the Laguna and the impact to floodwater storage. Flint and Flint (2012) downscaled regional climate models to estimate the effects of climate change on future water balance in the Russian River basin, and Flint et al. (2018) likewise used water balance modeling to examine scenarios for the impacts of drought in the watershed.

Prior Historical Ecology Analyses

The historical ecology mapping and analysis provided within this report builds upon several prior projects that reconstructed historical landscape characteristics for portions of the Laguna and its surrounding watershed. Butkus (2010, 2011b) developed historical and contemporary models to estimate nutrient loading within the Laguna watershed. Dawson and Sloop (2010) mapped the historical alignment of creeks in the southern headwaters of the Laguna, while Baumgarten et al. (2014) examined changes in the historical channel locations of Mark West Creek, the lower Laguna, and surrounding tributaries. *Historical Ecology and Landscape Change in the Central Laguna de Santa Rosa* reconstructed historical wetland and channel patterns for a central portion of the Laguna (Baumgarten et al. 2017).

Other Key Resources

Several additional documents provided background information on the Laguna's cultural and biotic resources, sediment and groundwater processes, and articulated conservation goals that informed those outlined in this report. Waaland (1989) and the Laguna Technical Advisory Committee (Laguna TAC 1989) provided general descriptions of the Laguna's history and ecology, and the Laguna TAC also listed broad goals for habitat restoration. The Recovery Strategy for California Coho Salmon (CDFG 2004) provided more specific restoration goals to support salmon in Santa Rosa and Mark West Creeks, whereas the Santa Rosa Plain Conservation Strategy (USFWS 2005) and Recovery Plan (USFWS 2016) outline conservation goals to benefit the California tiger salamander and federally listed endangered plants. The Santa Rosa Creeks Master Plan (City of Santa Rosa 2013) articulates goals for stream restoration, provides detailed descriptions of stream condition, and recommends management actions that will benefit the lower portions of tributaries to the Laguna and the Laguna itself. PWA (2004a; 2004b) provided two studies of sediment processes, including sources, rates, and flooding potential; while Winzler & Kelly GHD (2012) identified areas suitable for groundwater recharge and flood detention in the Laguna watershed. §

Riparian habitat and agricultural land near Laguna Wildlife Area Photo: Google Earth.

2 Laguna Restoration Vision Process

The Process for Developing a Vision

This Vision was developed through a collaborative science-based process focused on developing goals and identifying opportunities for restoring habitats and improving ecosystem functioning for wildlife and people in a manner that is resilient to current and future pressures. The development process included a detailed analysis of landscape change (both past and potential future), workshops with technical advisors and stakeholders to discuss findings and desired restoration approaches, and a synthesis of technical work and advisor and stakeholder feedback into a map of restoration opportunities and their associated ecosystem benefits. This approach, described in detail below, has been successfully applied for multi-benefit restoration planning efforts throughout the region (e.g., Lower Novato Creek, Lower Walnut Creek, Lower Calabazas and San Toma Aquino Creeks, and Upper Penitencia Creek).

UNDERSTAND LANDSCAPE FUNCTIONING

Vision development began by establishing an understanding of the landscape functioning from a past, present, and future perspective. This started with a synthesis of existing information on the key landscape-scale drivers for habitat support and ecosystem function, and how they have changed since the onset of intensive settlement. Next came the construction of a detailed picture of the historical ecology of the Laguna and its surrounding landscape, and the magnitude of change in habitat conditions over the past two centuries. This effort used previous historical ecology work to reconstruct historical landscape features and processes throughout the entire Laguna, and quantify change in selected key landscape features and processes. The final element was a general review of the anticipated future changes to landscape physical and ecological functioning associated with a changing climate.

OUTREACH AND WORKSHOPS

A series of meetings and workshops were held in 2018 and 2019 to gather expert knowledge on landscape history, function, and management processes, and to brainstorm and assess vision goals, objectives, and actions. On March 12, 2018, Sonoma Water, the Laguna de Santa Rosa Foundation, and SFEI organized a community meeting to share information about the project and the plan development process, and to answer questions and hear thoughts from the community. On December 5, 2018, a meeting of local landowners was held to discover landowner priorities and challenges for managing their land, and to solicit guidance on how to provide multiple benefits through restoration designs.

On March 13, 2019, the project Technical Advisory Committee (TAC) was convened for a workshop focused on identifying restoration opportunities and constraints, and technically feasible actions that could be included in the vision. The workshop began with presentations by SFEI describing a synthesized understanding of landscape evolution and functioning that was followed by a group mapping exercise to

The Laguna in flood, 1930. Photo: Sonoma Heritage Collection - Sonoma County Library.

identify priority restoration areas within the Laguna study area and in the wider Laguna watershed. TAC members included scientists and managers with expertise in hydrology, groundwater, plant ecology, fisheries, and wildlife ecology (for list of TAC members, see the Acknowlegements).

SFEI then applied TAC input to develop a preliminary vision map, which was presented as a draft to the project Management Advisory Committee (MAC) at a workshop held May 17, 2019. Members of the MAC included local regulatory and non-regulatory agency partners who operate within the Laguna and have expertise in land and water management. The goals of the workshop were to solicit MAC members' ideas about how to address management challenges in the Laguna, and to receive feedback on the vision goals, objectives, and initial vision map. MAC members contributed valuable suggestions that honed draft vision goals and objectives, as well as draft vision actions.

DEVELOP THE VISION

The restoration opportunities identified during the outreach and workshops were synthesized into two sets of landscape-scale measures. The first set emphasizes recommended actions within the current 100-year floodplain of the Laguna, and features specific areas where actions can be taken to improve ecological functioning within the Laguna. The second set focuses on recommended actions within the surrounding Laguna watershed that should be addressed to support within-Laguna actions. Both sets of measures are aimed at addressing landscape resilience by supporting ecosystem functions such as improved water quality, sediment delivery, flood protection, and habitat conditions for native wildlife in the Laguna and in the surrounding region.

Field visit in the Laguna with TAC members. Photo: SFEI.

TAC members discuss restoration options. Photo: SFEI.

LANDSCAPE RESILIENCE

This landscape visioning process was informed by the landscape resilience framework, which provides guidance on how to incorporate principles of landscape-scale ecological resilience into management actions. The framework identifies seven dimensions of landscape resilience, including: setting, process, connectivity, diversity/complexity, redundancy, scale, and people (Beller et al. 2015). Table 2-1 defines these dimensions and articulates recommendations for addressing them in the Laguna.

Landscape resilience is defined as "the ability of a landscape to sustain desired biodiversity and ecological functions over time in the face of climate change and other anthropogenic and natural stressors". ***Desired biodiversity*** is defined as including "native taxa, nearby species whose ranges may shift in the future, and nonnative species that support desired ecological functions or ecosystem services." ***Natural stressors*** include "both episodic events such as fire, flood, or drought, and prolonged stressors and directional change" (such as changes in precipitation, climate, or food availability) (Beller et al. 2019). Building landscape resilience would help improve ecological resilience to benefit people and wildlife in the Laguna, and help increase resilience to climate changes, both in the near and more distant future. §

Table 2-1. Assessment of Dimensions of Landscape Resilience in the Laguna de Santa Rosa using the Landscape Resilience Framework (after Beller et al. 2015).

Dimension	Definition	Recommendation
Setting	Unique geophysical, biological, and cultural aspects of a landscape that determine potential constraints and opportunities for resilience	Characterize setting and evaluate hydrologic, geomorphic, and biological processes in the Laguna. Addressed by: Changes in Key Drivers (Chapter 3), Historical Ecology (Chapter 4), Landscape Change Analysis (Chapter 5)
Process	Physical, biological, and chemical drivers, events, and processes that create and sustain landscapes over time	Evaluate processes and drivers that form the Laguna landscape in the past, present, and future. Addressed by: Key Drivers of Ecosystem Structure and Function (Chapter 3), Landscape Change Analysis (Chapter 5), Future Conditions (Chapter 6), Vision Concepts and Metrics (Chapter 7)
Connectivity	Linkages between habitats, processes, and populations that enable movement of materials and organisms	Protect and restore habitat quality and configuration to support beneficial outcomes for hydrologic, geomorphic, and biological processes (e.g., sediment management, nutrient cycling) and wildlife populations. Addressed by: Landscape Change Analysis (Chapter 5), Vision Concepts and Metrics (Chapter 7)
Diversity/ Complexity	Richness in the variety, distribution, and spatial configuration of landscape features that provide a range of options for species	Acknowledge and support the complexity of habitats in the Laguna when designing wetland and riparian restoration. Addressed by: Landscape Change Analysis (Chapter 5), Vision Concepts and Metrics (Chapter 7)
Redundancy	Multiple similar or overlapping elements of functions within a landscape that promote diversity and provide insurance against loss	Promote multiple large patches of contiguous habitat areas. Configure wetlands and riparian areas to support wildlife and filtration of sediment and pollutants. Addressed by: Vision Concepts and Metrics (Chapter 7)
Scale	The spatial extent and time frame at which landscapes operate that allows species, processes, and functions to persist	Support local, watershed-scale, and regional implementation of restoration concepts to benefit biodiversity and support ecosystem services. Addressed by: Vision and Watershed-Scale Concepts (Chapter 7)
People	The individuals, communities, and institutions that shape and steward landscapes	Encourage implementation of restoration projects through engaging landowners and the public via education, outreach, and incentives. Integrate recommendations into municipal and regional regulatory programs. Addressed by: Key Considerations (Chapter 8), Next Steps (Chapter 9)

Aerial view of the Laguna near Sebastopol. Photo: **Google Earth.**

3 Overview of Changes in Key Drivers of Ecosystem Structure and Function

Introduction

Developing an effective approach for restoring lost habitats within the Laguna starts with a solid understanding of the key physical processes that create and maintain habitats, and how those processes have changed since the onset of intensive development. As with all freshwater ecosystems, habitat conditions within the Laguna are controlled and supported in large part by the delivery of water (surface and subsurface), sediment, and nutrients from the surrounding watershed. This chapter provides a high-level overview of these dominant drivers of Laguna habitat conditions, and in turn, ecosystem structure and function, within the historical (ca. 1850s) and modern landscape. The information provided here is intended to help define changes in the key drivers over time, which will help inform management actions that support habitat restoration and long-term habitat resilience within the Laguna.

Changes in Hydrology

Surface and subsurface flows into the Laguna de Santa Rosa from the surrounding watershed originate in the Sonoma and Mayacamas Mountains to the east, and the Gold Ridge (Mendocino Range) to the west (Fig. 3-1).

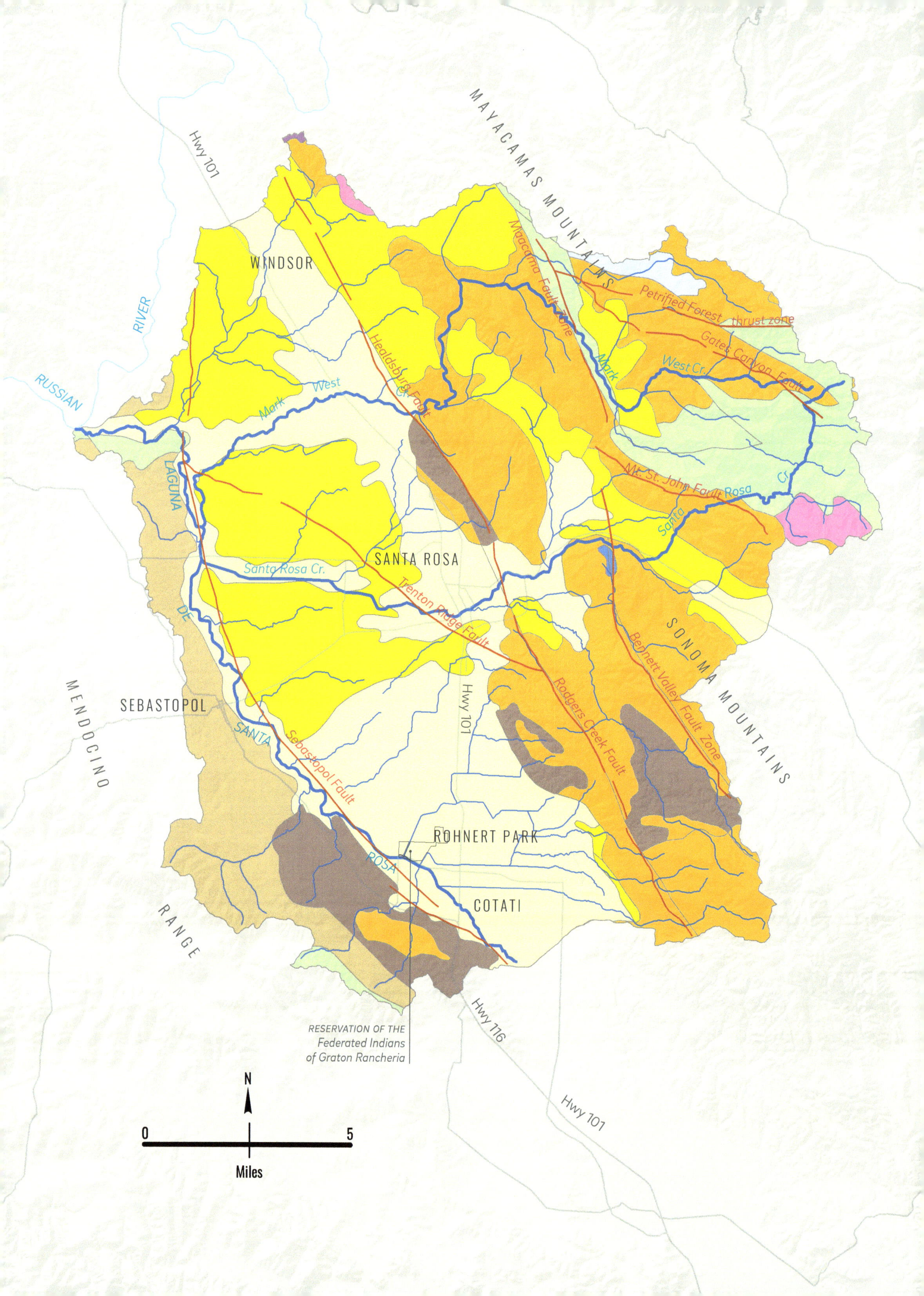

MAYACAMAS MOUNTAINS
Hwy 101
WINDSOR
RUSSIAN
RIVER
Mark West
Mark
West Cr.
Maacama Fault Zone
Petrified Forest thrust zone
Gates Canyon Fault
Healdsburg Fault
Mt. St. John Fault
Santa Rosa Cr.
LAGUNA
DE
SANTA
ROSA
SANTA ROSA
Santa Rosa Cr.
Trenton Ridge Fault
SONOMA MOUNTAINS
Bennett Valley Fault Zone
Rodgers Creek Fault
MENDOCINO
RANGE
SEBASTOPOL
Sebastopol Fault
ROHNERT PARK
COTATI
Hwy 116
RESERVATION OF THE
Federated Indians
of Graton Rancheria
N
0
5
Miles

Geologic Features

Melange

Franciscan Complex

Metavolcanic and Mixed

Volcanic and Metavolcanic

Ultramafic

Sedimentary

Marine Sediments

Petaluma Formation

Quaternary Deposits: Glen Ellen

Other Quaternary Deposits

Wilson Grove Formation

Volcanics

Sonoma Volcanics

Other

Water

Figure 3-1. (left) Geologic features of the Laguna de Santa Rosa watershed. Source: Nishikawa et al. 2013.

Historically, during light to moderate wintertime storm events, a large portion of precipitation was intercepted by the forests covering the mountains and infiltrated into the soil, with a modest amount of runoff and stream flow generation (Fig. 3-2). During large wintertime storm events, extensive runoff would lead to stream flows that would accumulate in steep headwater channels and then spread out onto broad alluvial fans at the base of the mountains before continuing downstream onto the Santa Rosa Plain and ultimately into the Laguna (Dawson and Sloop 2010). The inflow of the Russian River during these storm events prevented drainage and caused widespread flooding. Flow that infiltrated into the alluvial fans would recharge deep aquifers in the Sonoma Volcanics (andesite and basaltic tuffs) and Petaluma Formation (poorly sorted silty/clayey sand and gravels) to the east and the Wilson Grove Formation (well sorted marine sand) to the west (Nishikawa et al. 2013, Woolfenden and Nishikawa 2014). Flow from the east that was delivered further downstream onto the Santa Rosa plain would recharge the surficial Glen Ellen Formation (Quaternary alluvium). Wintertime storm flow that was delivered into the Laguna would cause flooding that typically lasted late into the spring when the receding Russian River stage would allow the Laguna to drain. During the dry season, groundwater levels were typically within several feet of the ground surface. Groundwater would discharge into the lowest reaches of some tributaries, providing surface flows into the Laguna, and upwell directly into the Laguna (Nishikawa et al. 2013).

Over the past 200 years since the onset of intensive European American settlement, widespread landscape modification has had considerable impacts on the flooding dynamics within the Laguna. Within the Laguna's surrounding watershed, land use and channel changes have resulted in an overall increase in runoff volume and decrease in the travel time

HISTORICAL

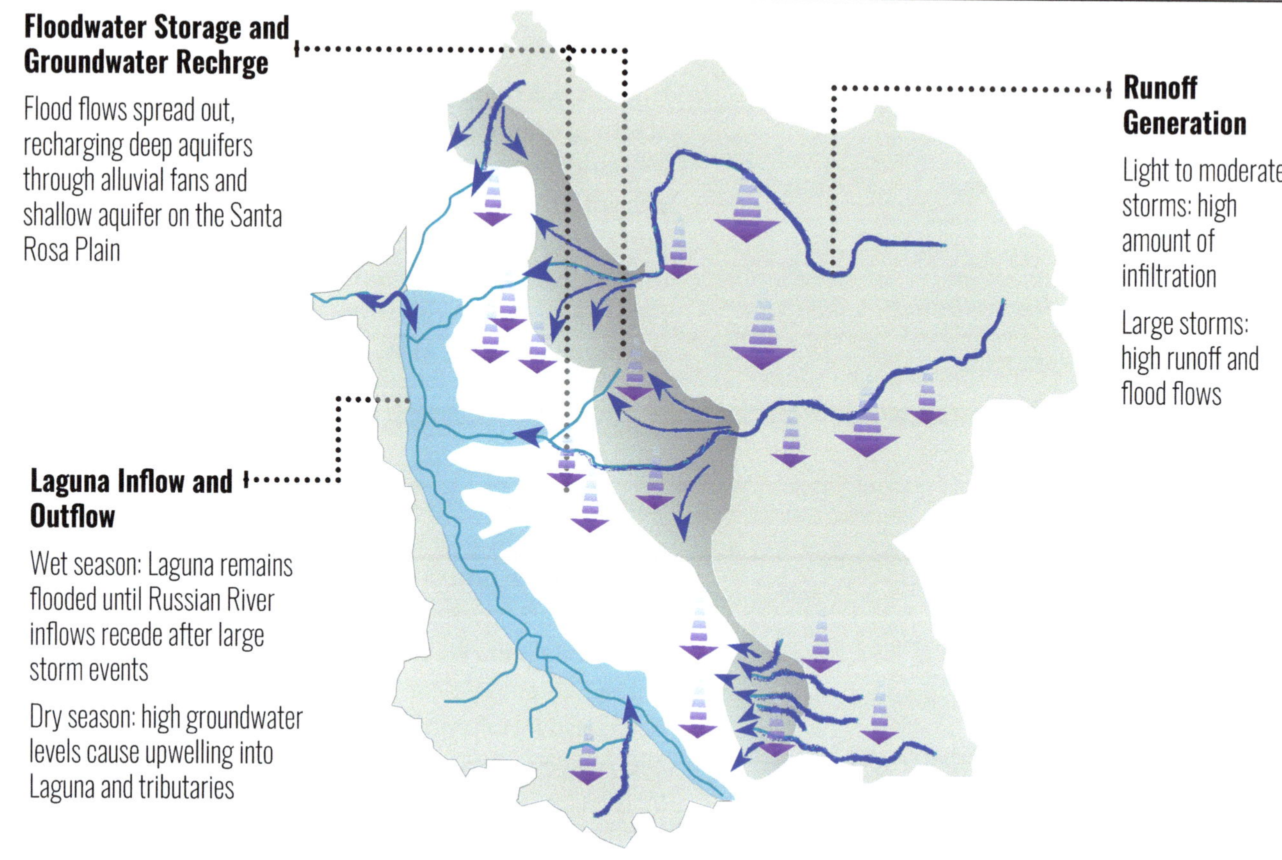

MODERN

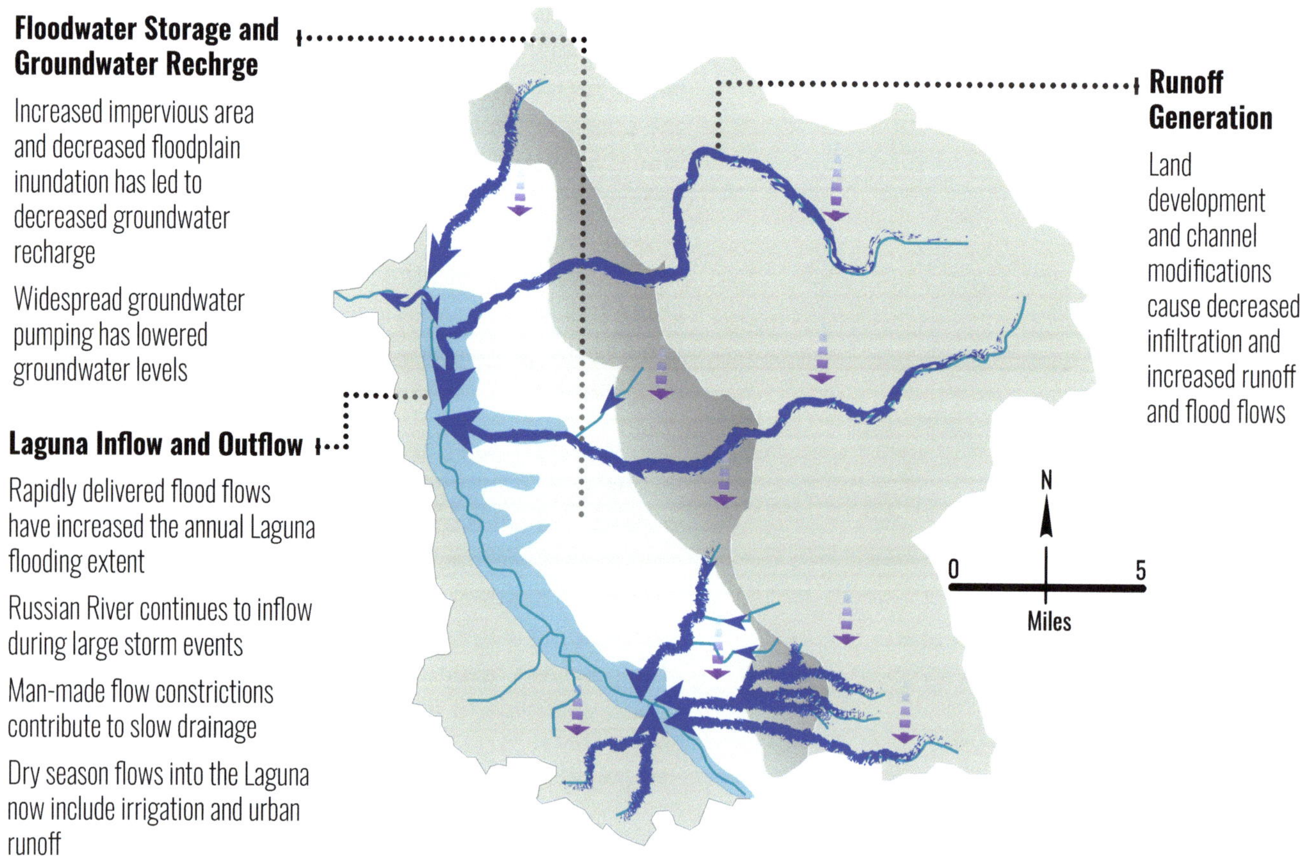

Conceptual Model Features

River Flow

Infiltration and Groundwater Recharge

Russian River Backwater Effect During High Flow

Mountain

Alluvial Fan

Santa Rosa Plain

Floodplain

Figure 3-2. (left) Conceptual model indicating past and present hydrologic conditions in the Laguna de Santa Rosa and its contributing watershed.

for flood flows entering the Laguna (PWA 2004b, Sloop et al. 2007, Curtis et al. 2013). Within the headwaters of Laguna tributaries, the loss of forests, and building of impervious surfaces and road networks has decreased infiltration during storm events and increased the amount of runoff and streamflow (Sloop et al. 2007). On the alluvial fans at the base of the mountains and on the Santa Rosa Plain, the conversion of forested land and wetlands to agricultural and urbanized lands has further decreased infiltration and floodwater storage, and increased runoff. In addition, the straightening, lengthening, rerouting, and leveeing of tributary channels for flood management has resulted in the rapid delivery of flood flows to the Laguna (Curtis et al. 2013). Within the Laguna itself, flow constrictions caused by levees and bridge crossings contribute to relatively slow drainage during and following the wet season. This is a major issue in the Guerneville Road Bridge area, where flow constriction is exacerbated by Delta Pond levees (E. Andrews, pers comm; B. Cluer, pers comm; L. Flint, pers comm). Anecdotal observations suggest the annual flooding extent in the Laguna is expanding, which is attributed to the combination of rapid delivery of stormwater and local backwater effects that prevent drainage (Curtis et al. 2013) (Fig. 3-3, page 29).

Like flood flows, groundwater and dry season flows have changed considerably over the last 200 years. In general, groundwater levels have decreased compared to historical conditions due to a combination of decreased stormwater infiltration and widespread groundwater pumping (Sloop et al. 2007, Nishikawa et al. 2013, Woolfenden and Nishikawa 2014). Impervious surfaces on the alluvial fans and Santa Rosa Plain prevent stormwater infiltration and deliver runoff rapidly to adjacent channels. Many of the channels that drain these urbanized landscapes are now culverted or lined with concrete, which further decreases the opportunity for stormwater infiltration. Groundwater pumping in the region began in the 1870s, and by the early 1950s, the Santa Rosa Plain had approximately 8,500 groundwater wells (Cardwell 1958). Between 1974-2009, the average total annual pumpage from all wells was approximately 47,400 acre-feet of water (Nishikawa et al. 2013). During this time period, groundwater pumping caused springtime groundwater levels to drop 20 ft or more in many locations throughout the Santa Rosa Plain, causing tributary channels to lose more water to the groundwater system than they gained (Nishikawa et al. 2013, Woolfenden and Nishikawa 2014). Currently, dry season flow that enters the Laguna from tributaries is derived from a combination of groundwater discharge, irrigation runoff, and urban runoff (Nishikawa et al. 2013).

The Laguna in flood. Photo: SFEI.

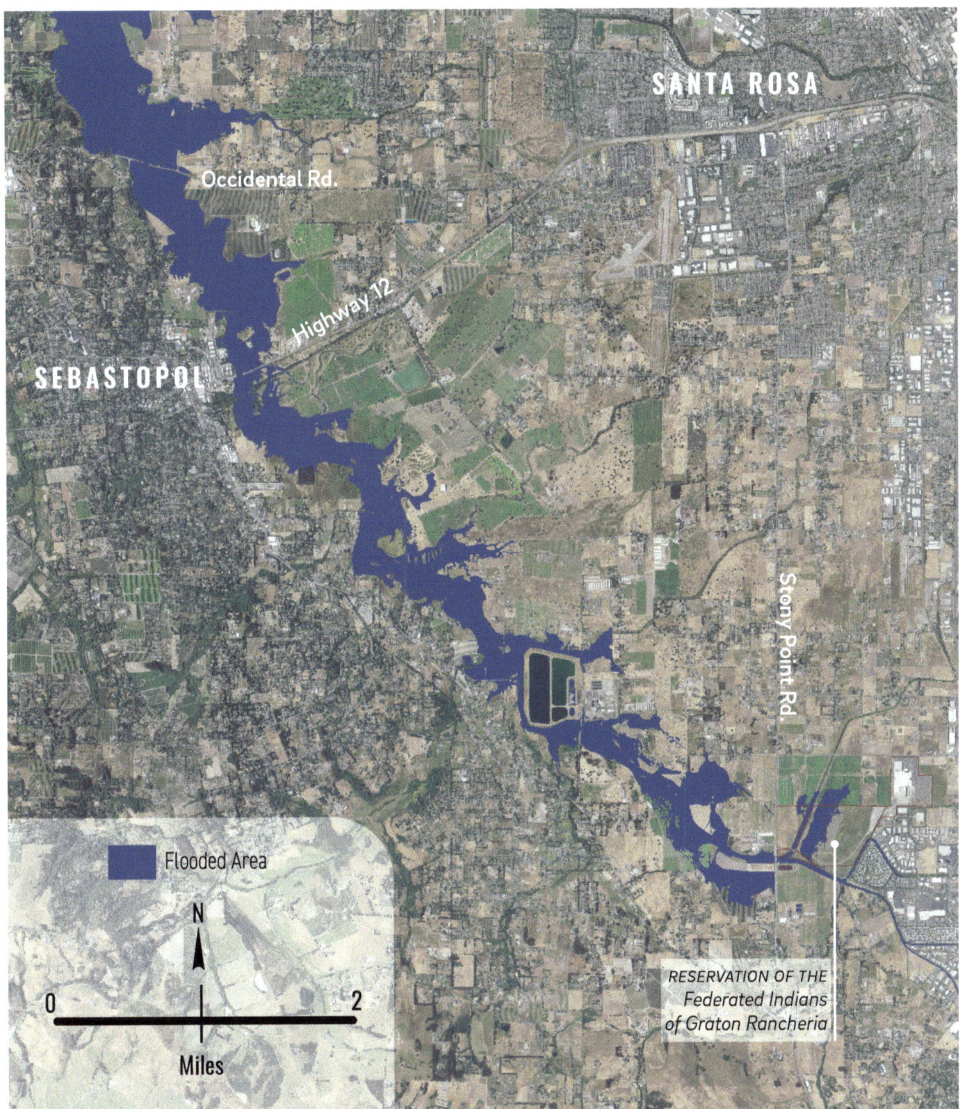

Figure 3-3. Flooded area in the southern portion of the Laguna for a typical wintertime storm event based on local stage data (January 20, 2010). Source: Nathan Baskett, Sonoma Water; USGS gage on Laguna de Santa Rosa Creek near Sebastopol (USGS 1465750).

Changes in Sediment Dynamics

The geologic and hydroclimatic setting of the Laguna and its surrounding watershed result in regionally high rates of both watershed sediment production and sediment storage within and adjacent to the Laguna (Fig. 3-4). The mountain ranges that flank the Laguna are tilting blocks of bedrock: the Santa Rosa block underlying the Sonoma and Mayacamas Mountains to the east, and the Sebastopol Block underlying the Gold Ridge to the west (PWA 2004b). The Santa Rosa Plain and the Laguna are in a depressional sedimentary basin atop the Windsor Syncline that is subsiding relative to the adjacent uplifting blocks (Curtis et al. 2013). The Rodgers Creek fault marks the break between the Santa Rosa Plain and the uplifting Santa Rosa Block to the east and the Sebastopol fault marks that break between the Valley and uplifting Sebastopol Block to the west (Fig. 3-1). Upstream of these faults on both blocks, the main Laguna tributaries have rapidly eroding v-shaped valleys, many with steep landslide prone valley walls, resulting in a relatively high natural rate of fine and coarse sediment production (PWA 2004b). This, combined with average annual precipitation between 45-55 inches at the highest elevations (PRISM Climate Group 2019), results in inherently high supply downstream. Coarser sediment coming from the main tributaries during large storm events would historically deposit on alluvial fans and build them out over time. Finer sediment was transported further downstream where it would be deposited on the Santa Rosa Plain and in the Laguna as the landscape slope decreased and flood flows lost power. Within the Laguna, backwater caused by high Russian River stage would cause finer sediment to deposit, particularly downstream of the Santa Rosa Creek confluence (Curtis et al. 2013). Consequently, a large percentage of sediment produced during storm events would remain stored within and upstream of the Laguna.

As with hydrology, landscape modifications over the past 200 years have considerably altered Laguna sediment dynamics (Fig. 3-4). In the upper watershed, land clearing and the building of impervious surfaces and road networks has led to increased mass wasting, channel erosion, and

Conceptual Model Features

Erosion

Sediment Transport

FIne Sediment Deposition

Mountain

Alluvial Fan

Santa Rosa Plain

Floodplain

Figure 3-4. (right) Conceptual model indicating past and present sediment dynamics in the Laguna de Santa Rosa and its contributing watershed.

HISTORICAL

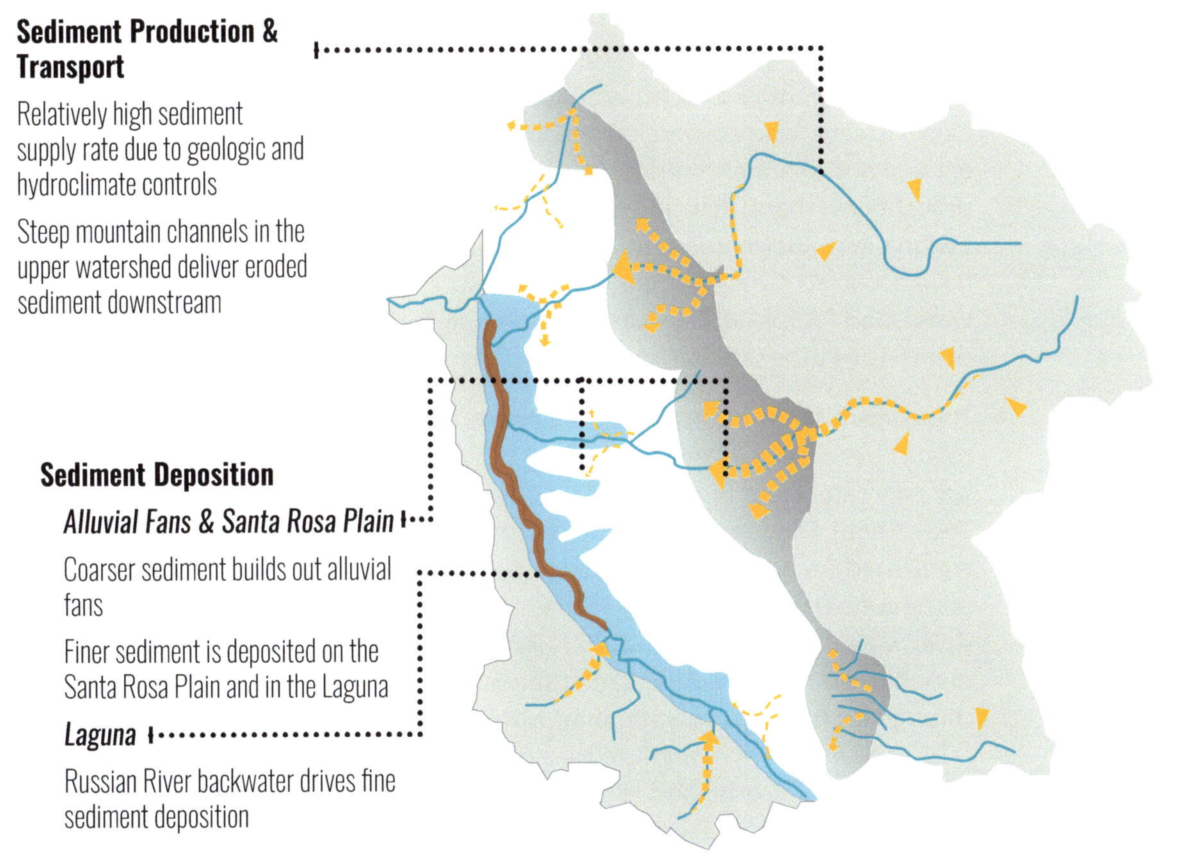

MODERN

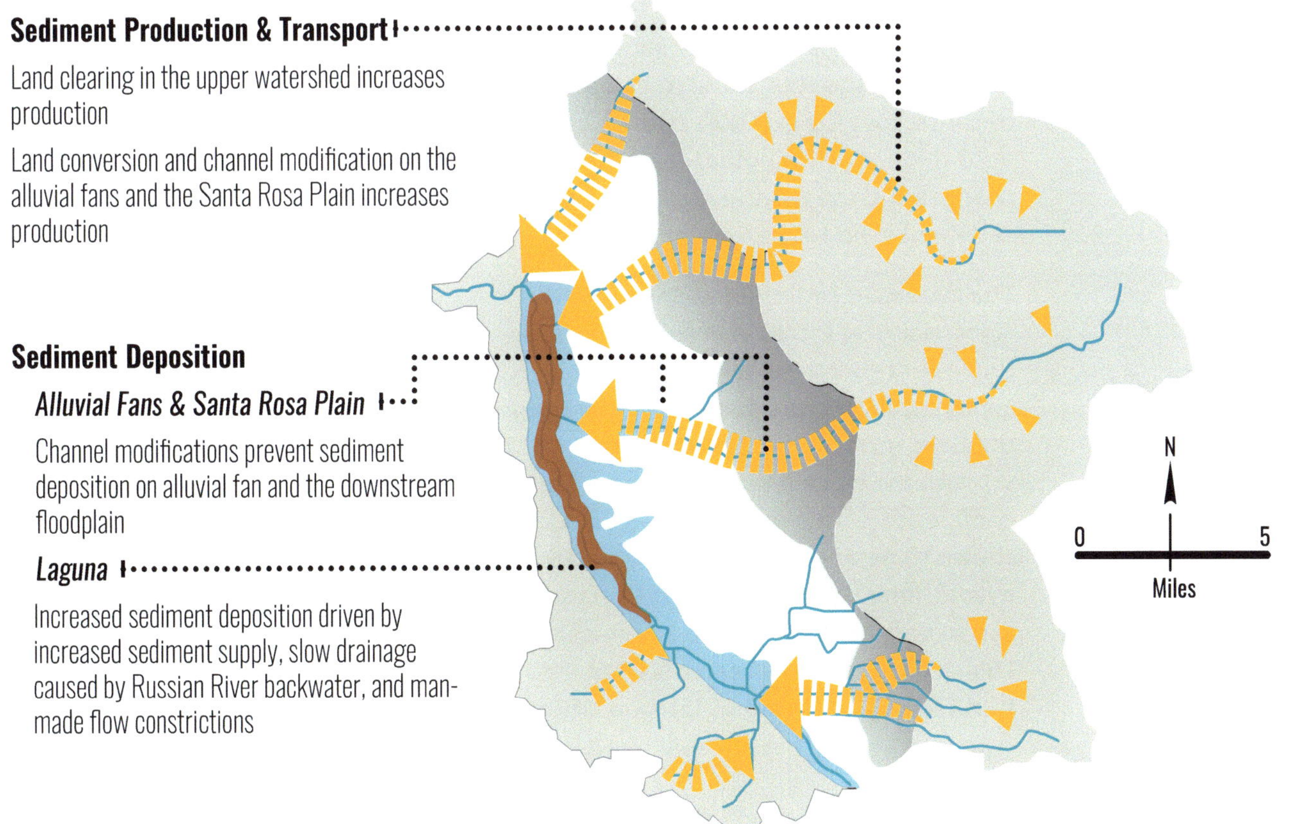

overall sediment yield compared to historical conditions (Sloop et al. 2007). Downstream, land conversion and channelization have increased surface erosion and channel erosion, while channel leveeing for flood management prevents overbank flow and directs fine sediment downstream to the Laguna. An initial increase in sediment supply to the Laguna is thought to have occurred during the period of widespread agricultural production in the watershed (1850s-1950s), with an additional increase occurring with the onset of intensive urbanization (post-1950) (PWA 2004b).

The current sediment yield to the Laguna is estimated to be 120-490 tonnes per square kilometer per year ($t/km^2/yr$), which is thought to be approximately four to ten times pre-development values (PWA 2004b, Tetra Tech 2015b). For comparison, the sediment yield for the entire Russian River watershed from 1939-2005 is estimated to be 318 $t/km^2/yr$ (Wheatcroft and Sommerfield 2005) and the sediment yield for the neighboring Petaluma River watershed from 1995-2016 is estimated to be 385 $t/km^2/yr$ (L. McKee, unpublished data). Widespread cropland, vineyard, and developed areas throughout the Laguna watershed with locally high sediment yield rates (>350 $t/km^2/yr$) are thought to be a primary cause for the high current sediment yield (Potter and Hiatt 2009, Tetra Tech 2015b). For example, the high occurrence of these areas in the Upper Laguna, Lower Floodplain, and Windsor Creek subwatersheds helps drive a relatively high sediment supply to the Laguna relative to their contributing area (Tetra Tech 2015b). Channel erosion is also thought to contribute to the increased sediment yield, particularly within the Windsor Creek, Mark West Creek, and Blucher Creek subwatersheds (PWA 2004b). Field-based channel incision estimates range from 0.9-1.2 m (3-4 ft) in the alluvial fan section of Copeland Creek (Laurel Marcus & Associates 2004) to at least 1.8 m (6 ft) in the alluvial fan section of Santa Rosa Creek (PWA 2004b). However, more information is needed to determine which tributary reaches are still eroding in response to landscape changes and the relative contribution of channel erosion to the current sediment yield to the Laguna.

The increased sediment supply from the Laguna watershed has led to a corresponding increase in sediment deposition within the Laguna and the lower tributary reaches (Sloop et al. 2007). Over the past 50+ years, sedimentation rates throughout the Laguna are estimated to have almost doubled, increasing from an average of 2.7 mm/yr to 4.5 mm/yr (0.1-0.2 in/yr), with the highest rates still occurring downstream of the Santa Rosa Creek confluence (Curtis et al. 2013). Recent local average annual rates have been shown to be as high as 14.5 mm/yr (0.6 in/yr; Aalto 2004). During large storm events, finer sediment deposits up to 25 cm (9.8 in/yr) thick have been noted at tributary confluences, and coarser alluvial fan sediment deposits up to 1.5 m (5 ft) thick have been noted at the mouths of the steep western tributaries (Curtis et al. 2013). Approximately one quarter of the sediment deposited in the mainstem Laguna channel and the lower reaches of Laguna tributaries is now removed to improve flood conveyance (Tetra Tech 2020). Between 2008-2014, Sonoma Water removed approximately 20,000 tonnes/yr from the Laguna mainstem and tributary channels, with the sediment removal "hotspots" being lower Copeland Creek, lower Hinebaugh Creek, lower Gossage Creek, mainstem Laguna from Hinebaugh Creek to Colgan Creek confluence, lower Colgan Creek, and lower Santa Rosa Creek (Tetra Tech 2015b; Sonoma Water, unpublished data). The estimated mass of deposited watershed sediment that currently remains in the Laguna after sediment removal (~55,000 tonnes/yr) is thought to be over ten times the historical amount of watershed sediment that deposited in the Laguna on an average annual basis (Tetra Tech 2020).

Landslide on upper Copeland Creek (circled person is approximately six feet tall). Photo: SFEI.

Changes in Nutrient Dynamics

The Laguna was likely a relatively productive system historically, as evidenced by historical records of its abundant populations of 'salmon-trout', thousands of visiting waterfowl every winter, as well as accounts of plentiful beaver, elk, and pronghorn (Baumgarten et al. 2017). However, evidence of nutrient-poor soil types in some areas, and historical records of aquatic plants that thrive in nutrient-poor waters, points to a heterogeneous distribution of nutrients historically (Baye 2018). Estimates of the historical nutrient load coming into the Laguna indicate that they were small compared to modern conditions, since anthropogenic additions of nutrients were negligible before the 20th century (Waaland 1989, Butkus 2011b, Tetra Tech 2015a). Habitats within the Laguna were supported by the delivery of nutrients from the surrounding watershed in both surface water and groundwater; however, historically, the delivery of nutrients to the Laguna was relatively modest due to capture and storage of surface water and sediment upstream of the Laguna, and the removal of nutrients within groundwater as it flowed under the Santa Rosa Plain towards the Laguna. Within the Laguna, wetland complexes and riparian areas contributed to the interception, internal processing, and transformation of the nutrients that did arrive there, as waters slowly coursed through the system, resulting in a relatively small load discharging from the Laguna to the Russian River (Baumgarten et al. 2017).

An adequate supply of the essential nutrients nitrogen (N) and phosphorous (P) is necessary for maintaining production at the base of the food web. However, like many aquatic systems worldwide, Laguna waterways now experience an 'increase in the supply of organic matter to an ecosystem' known as eutrophication (Nixon 1995, 2009, Cloern 2001, Smith 2003). There are several drivers of eutrophication in the Laguna landscape.

Increased nutrient inputs and more efficient delivery to the Laguna. Nutrient inputs to the Laguna have greatly increased since the mid-19th century, with increased loads of N and P released in surface runoff from cities, farms, and ranches. In the Laguna, instream concentrations of total N have seen a net long-term increase, and exceed federal and state criteria. While P has been reduced from peaks in the 1980s and 1990s, it remains at concerning levels in the Laguna (Fitzgerald 2013). The delivery of these nutrients has increased via alterations of physical habitat and hydrology, mainly due to increased agricultural and urban land cover and decreases in riparian and wetland areas known to trap and process nutrients. Additionally, increased channel length and connectedness, and increased area of impervious surfaces, contribute to the increased delivery of nutrient-laden water and sediments to the Laguna. A portion of this nutrient-rich runoff also percolates to groundwater.

Increased storage, internal cycling, and export of nutrients. The Laguna has responded to the elevated loads of N and P through increased storage, recycling, and processing of these nutrients. Some of the nutrients have been buried under successive layers of sediments, and can become remobilized by bacteria at the sediment-water interface, or when the sediments are mixed or disturbed. N and P available in the water cycles through aquatic algae, plants, and animals within the Laguna, which take up nutrients during the growing season, then release them as they scenesce or die. Increased temperature and light availability in unshaded channels contribute to increased algal and aquatic plant growth during this recycling process. In the case of N, a fraction can be removed from water and sediments to the air through chemical transformations; P cannot be transformed in this way. Nutrients that are not buried or used for in-situ algae, plant, and animal growth are exported from the system along with flows, discharging to the Russian River and the Russian River Estuary.

Together these 'biostimulatory conditions' have increased the proliferation of undesirable aquatic algae and plants in the Laguna, including invasive water primroses (*Ludwigia* spp.) (Tetra Tech 2015b, Sutula et al. 2018). Additionally, increases in harmful algal blooms, increased bacterial growth, and warmer water temperatures reduce water quality for agricultural uses (Sutula et al. 2018). Eutrophic conditions in the Laguna also cause low dissolved oxygen and increased suspended solids that alter instream habitat conditions and shift plant and animal communities in ways that reduce habitat quality for native fishes. Increases in algae and floating aquatic plants associated with eutrophic conditions interfere with flood management and mosquito abatement, and reduce aesthetic and recreational value in the Laguna (Honton and Sears 2006).

Problems related to elevated concentrations and loads of nutrients and sediments in the Laguna have prompted regulatory and policy actions to reduce biostimulatory conditions in the watershed (Morris 1995, Fitzgerald 2013, Kieser & Assoc. 2015, NCRWQCB 2018). As part of its effort to determine a TMDL for N and P, the NCRWQCB (Butkus 2011b, see also Tetra Tech 2015a) developed a Land Cover Loading Model (LCLM) to estimate historical and modern nutrient loading in the Laguna watershed, finding a 3-fold increase in N and a 4.5-fold increase in P loading compared to historical conditions (Fig. 3-5, next page).

Landscape patterns of N and P loading become apparent from mapping the nutrient loading rates from the LCLM (see Figs. 3-6 and 3-7, pages 38-39). With the exception of contributions from wastewater treatment facilities, most loads to the Laguna are from non-point sources such as urban and agricultural runoff. N loads are greatest in urban and commercial areas, followed by cropland and pastures. Greatest N loads occur outside of the

Figure 3-5. **Median annual Total Phosphorous and Total Nitrogen loads (lb/yr) in the Laguna.** The LCLM (Butkus 2010) calculated N and P loads in several ways, including the median and mean total P and total N by land cover type, as well as loads for various forms of N and P. For purposes of historical comparison, this chart shows the median values for total N and total P.

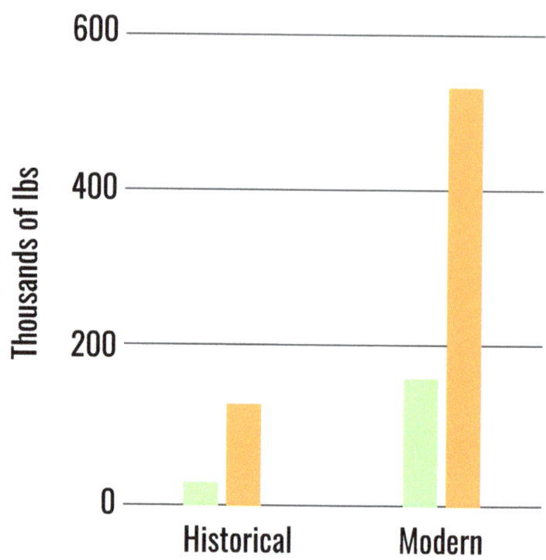

Laguna's 100-year floodplain. Phosphorous loads are highest in croplands and pastures, followed by residential areas and orchards; these land uses are located both within and outside of the 100-year floodplain.

Reducing both the loading and the concentrations of N and P will be needed to address eutrophication (Elser et al. 1990, Tetra Tech 2015a, Dodds and Smith 2016). One way managers try to reduce excess algae and plant growth is to determine the 'limiting nutrient' within a waterbody and to control it. A limiting nutrient is any chemical required for plant growth, but that is available only in small quantities. Once plants and algae consume the limiting nutrient, their populations stop expanding. For example, experiments in lake systems have shown reductions in plant growth when reducing only P (e.g., Schindler et al. 2008). However, in river and estuary systems like the Laguna, merely reducing P without also reducing N can result in continued undesirable algae growth (Taylor et al. 2004). Additionally, controlling only P does not address export of excess N to downstream areas, which can contribute to N-driven eutrophication in receiving waters and estuaries (Paerl 2009). Estimates of N and P concentrations in the Laguna suggest that both far exceed the limiting concentrations for plant growth, with concentrations of about 450 ug/L for inorganic N and 900 ug/L for inorganic P. Estimated limiting concentrations are 80 ug/L for N, and 20 ug/L for P (Sloop et al. 2007). Even when contributions of N and P from the surrounding watershed are reduced, it is possible that neither N nor P can become limiting in the Laguna because of releases of these nutrients from Laguna sediments. This indicates that both loads from the surrounding landscape, as well as legacy concentrations, must be reduced to have an impact.

Effective nutrient management will require efforts to reduce concentrations and loads across the landscape. Reducing both N and P in the Laguna and surrounding watershed can not only benefit local water quality, but also can contribute to beneficial reductions of nutrients in the Russian River and Russian River Estuary. §

Oak tree in the Laguna. Photo: SFEI.

HISTORICAL PATTERNS: Nitrogen Loading

Figure 3-6. Landscape patterns of Nitrogen loading in the Laguna watershed. Loading of Nitrogen has increased dramatically from historical conditions, with urban and residential areas located across the Santa Rosa Plain contributing the highest loads. Data source: Butkus 2011b; see also Tetra Tech 2015a.

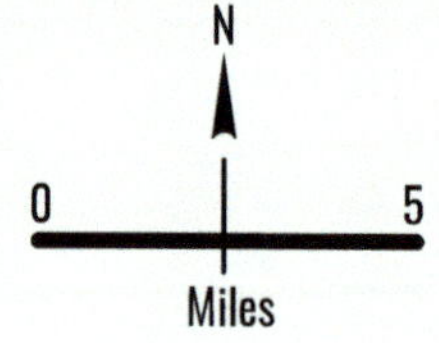

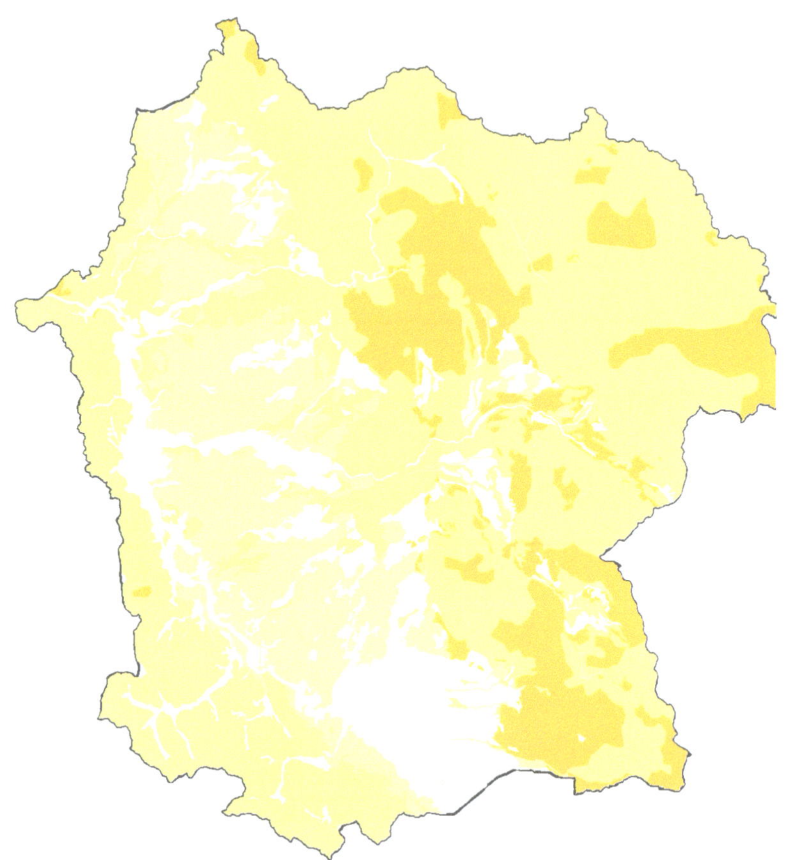

Historical Median Nitrogen Rate (lbs/ac/yr)

- 0 (Wetland, Riparian, Open Water)
- 0.41 (Oak Savanna)
- 0.73 (Forest)
- 2.64 (Rangeland)

MODERN PATTERNS: Nitrogen Loading

Modern Median Nitrogen Rate (lbs/ac/yr)

- 0.73 (Forest)
- 2.64 (Rangeland)
- 3.16 (Orchards and Vineyards)
- 4.11 (Other)
- 4.29 (Cropland and Pasture)
- 5.81 (Sewered Residential)
- 5.85 (Septic Residential)
- 6.29 (Commercial)

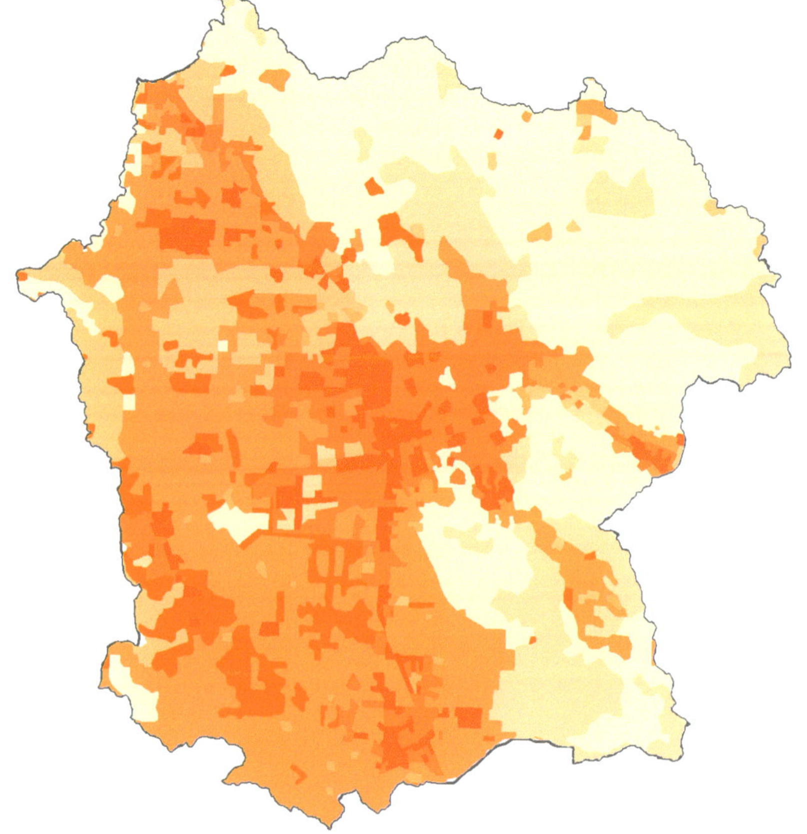

HISTORICAL PATTERNS: Phosphorous Loading

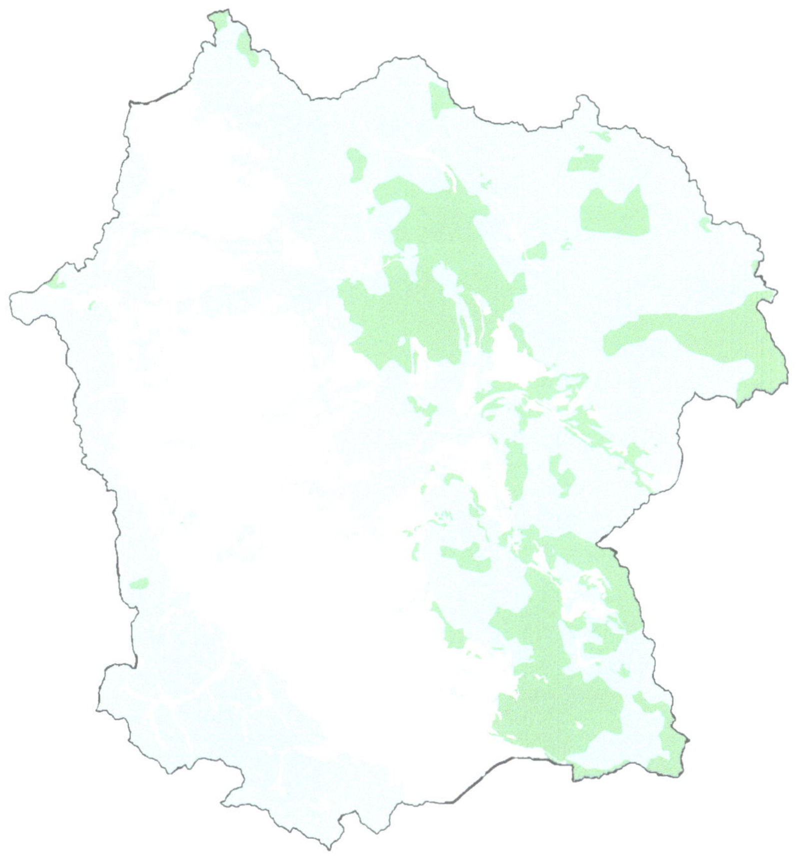

Figure 3-7. Landscape patterns of Phosphorus loading in the Laguna watershed. Phosphorus loading in the Laguna watershed has increased dramatically from historical conditions, with agricultural and residential areas located close to the Laguna contributing the highest loads on the modern landscape. Data source: Butkus 2011b; see also Tetra Tech 2015a.

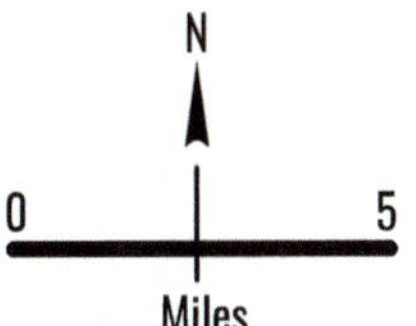

Historical Median Phosphorous Rate (lbs/ac/yr)

- 0 (Wetland, Riparian, Open Water)
- 0.09 (Oak Savanna)
- 0.12 (Forest)
- 0.72 (Rangeland)

MODERN PATTERNS: Phosphorous Loading

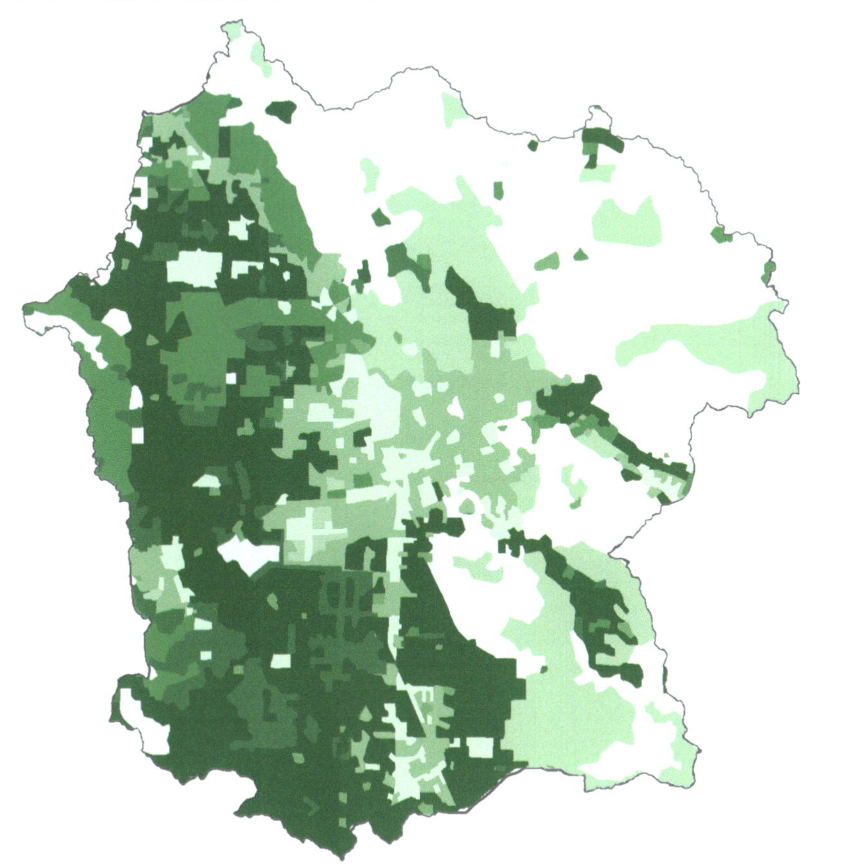

Modern Median Phosphorous Rate (lbs/ac/yr)

- 0.62 (Forest)
- 2.55 (Commercial)
- 3.32 (Rangeland)
- 3.47 (Residential with Sewer)
- 4.86 (Other)
- 5.89 (Orchards and Vineyards)
- 6.0 (Residential with Septic)
- 12.19 (Cropland and Pasture)

Boating on Lake Jonive, ca. 1907. **Photo:** CHS45682, courtesy of USC Libraries, California Historical Society Collection.

4 Historical Ecology

of the Laguna de Santa Rosa

Introduction

Examination of the historical ecology of the Laguna—which plants and animals it supported, how habitats were distributed along physical gradients, how water and sediment moved through the landscape—provides baseline information about the processes that formed and sustained specific Laguna habitats and the ecological functions these habitats provided. It is not intended to provide a prescription for how to restore the Laguna to a past condition. Rather, it provides crucial context for understanding how the Laguna has changed over time, which ecological functions have been lost, and how landscape changes have contributed to the management challenges that the Laguna faces today. When combined with contemporary research and projections of future changes, historical ecology provides an important tool to help identify appropriate restoration targets and develop a future vision for a healthy, resilient, and biodiverse Laguna landscape.

This chapter describes the methodology used to investigate the historical ecology of the Laguna and the results of this research. Later, the historical and modern Laguna ecology are compared to quantify the magnitude of habitat change over the past two centuries (Chapter 5).

Methods

OVERVIEW

Constructing an accurate picture of historical landscape patterns requires the integration, comparison, and interpretation of many independent sources (Grossinger et al. 2007). Where possible, historical landscape features in the Laguna were documented using multiple sources from varying years and authors to ensure accurate interpretation. This section details how these sources were collected and interpreted, as well as how they were used to create maps of historical habitat types and channels.

DATA COLLECTION AND COMPILATION

Historical data, including maps, photographs, and textual documents were collected from 24 local, regional, and state archives, as well as approximately 30 online databases (Table 4-1). The assembled dataset is composed of a variety of source types, including maps (e.g., land grant case maps, General Land Office (GLO) surveys, topographic maps, United States Department of Agriculture soil maps, county maps, parcel maps, railroad maps), photographs (landscape and aerial), and textual documents (e.g., land grant case files, General Land Office field notes, travelogues, newspaper articles, county histories, specimen records, and technical reports). In total, the dataset includes approximately 600 photographs, 550 maps, and 200 textual documents. High-value spatial data was geolocated in a Geographic Information Systems (GIS) database.

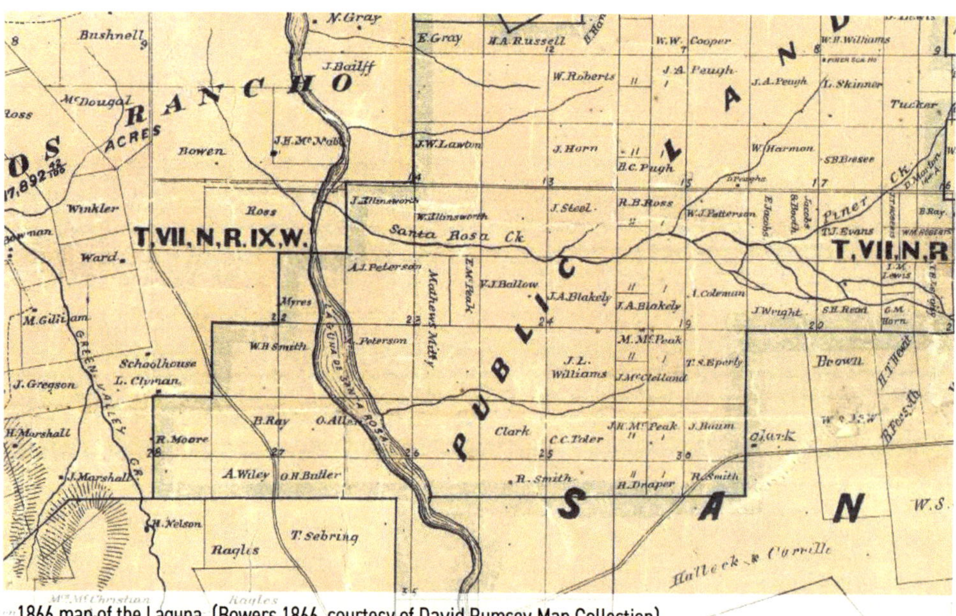

1866 map of the Laguna. (Bowers 1866, courtesy of David Rumsey Map Collection).

Table 4-1. Source institutions visited when collecting historical data.

Institution	Location
Local Archives	
Cotati Historical Society	Cotati
Curtis & Associates, Inc.	Healdsburg
Northwestern Pacific Railroad Historical Society	Petaluma
Sonoma County History and Genealogy Annex	Santa Rosa
Sonoma State University Library	Rohnert Park
Western Sonoma County Historical Society	Sebastopol
Laguna de Santa Rosa Foundation*	Santa Rosa
Petaluma History Museum*	Petaluma
Petaluma History Room, Sonoma County Library	Petaluma
County Archives	
Sonoma County Water Agency*	Santa Rosa
Sonoma County Surveyor, Department of Permit and Resource Management	Santa Rosa
Sonoma County Surveyor, Department of Transportation and Public Works	Santa Rosa
Sonoma County Recorder/Assessor	Santa Rosa
North Coast Regional Water Quality Control Board*	Santa Rosa
Regional Archives	
California Historical Society	San Francisco
Jepson Herbarium, UC Berkeley	Berkeley
Society of California Pioneers	San Francisco
The Bancroft Library, UC Berkeley	Berkeley
Earth Sciences and Map Library, UC Berkeley	Berkeley
The George & Mary Foster Anthropology Library, UC Berkeley	Berkeley
Sacramento Archives	
California State Archives	Sacramento
California State Library	Sacramento
California State Railroad Museum Library	Sacramento
Bureau of Land Management	Sacramento

* Denotes institutions visited for previous SFEI historical ecology studies within the study area.

SYNTHESIS AND MAPPING

Historical documents and contemporary spatial data were synthesized in a GIS database to develop a digital map that represents historical habitat and channel configuration ca. 1850. Within the 3,380 ha mapping extent, habitats were mapped as polygons, and were classified into 12 habitat types: Perennial Freshwater Lake/Pond, Valley Freshwater Marsh, Willow Forested Wetland, Mixed Riparian Forest, Wet Meadow, Seasonal Lake, Vernal Pool Complex, Valley Grassland, Oak Savanna, Oak Savanna/Vernal Pool Complex, Oak Woodland, or Mixed Conifer Forest. Table 4-2 (facing page) describes each of these habitat types according to their vegetation component characteristics: hydrology/flooding regime, and soil/drainage characteristics.

Historical channels were mapped as line features, and were classified as either Channel (well-defined drainage features), Slough (shallower, low-gradient drainage features through wetland complexes), or Side Channel (infrequently activated channels in a multi-threaded channel network).

Wetland and channel features were mapped (or "digitized") from the most spatially accurate sources believed to be representative of historical landscape conditions and configurations. The North Coast Aquatic Resource Inventory (NCARI) stream network data layer (SFEI-ASC 2014) was used as the starting place for digitizing channels. NCARI mapping was retained where it was found to reflect channel location depicted in historical sources and modified where evidence suggested that channel location had been altered. In many cases, historical features were digitized from a combination of historical maps, the 1942 aerial photographs (USDA 1942), and the modern Light Detection and Ranging (LiDAR)-derived digital elevation model (DEM) (WSI 2013). Wherever possible, early sources (e.g., mid-late 19th century maps and textual data) were used to confirm the historical presence of a particular feature and establish its approximate shape, size, location, and classification. Mapped features were attributed with both digitizing and "interpretation" (supporting) sources, as well as certainty levels for shape, location, and interpretation (classification). See Table 4-3 (below) for a definition of the certainty levels assigned to each feature.

Table 4-3. Certainty levels assigned to each of the features in the Laguna de Santa Rosa historical synthesis mapping. (Certainty levels are included as GIS attributes, and are not displayed here.)

Certainty Level	Interpretation	Size	Location
High/"Definite"	Feature definitely present before Euro-American modification	Mapped feature ~90-110% of actual feature size	Expected max. horizontal displacement less than 50 m (150 ft)
Medium/"Probable"	Feature probably present before Euro-American modification	Mapped feature ~50-200% of actual feature size	Expected max. horizontal displacement less than 150 m (500 ft)
Low/"Possible"	Feature possibly present before Euro-American modification	Mapped feature ~25-400% of actual feature size	Expected max. horizontal displacement less than 500 m (1,600 ft)

Table 4-2. Habitat Types.

Habitat Type	Characteristic Plants	Hydrology/Flood Regime	Soil/Drainage Characteristics
Valley Freshwater Marsh	Tules and bulrushes (*Bolboschoenus* and *Schoenoplectus* spp.), cattails (*Typha* spp.), sedges (*Carex* spp.), and rushes (*Juncus* spp.).	Semi-permanently to perennially flooded; subsurface perennially saturated	Poorly drained clay and clay loam soils
Willow Forested Wetland	Willows (*Salix* spp.), Oregon ash (*Fraxinus latifolia*), tule (*Schoenoplectus* spp.)	Seasonally to semi-permanently flooded; subsurface perennially saturated	Poorly to well drained soils; variable texture
Perennial Freshwater Lake/Pond	Submersed aquatic plants (e.g., *Potamogeton* spp.), open water	Perennially inundated	Inundated; mud substrate
Mixed Riparian Forest	Oaks (*Quercus* spp.), willows (*Salix* spp.), Oregon ash (*Fraxinus latifolia*), maples (*Acer* spp.)	Temporarily flooded from surface runoff; access to shallow groundwater in the hyporheic zone	Mixed alluvium, poorly to well drained soils (lower levels of clay in upper horizons than marshes/wetlands)
Wet Meadow	Grasses (e.g., *Elymus triticoides*), rushes (*Juncus* spp.), spike-rushes (*Eleocharis* spp.), sedges (*Carex* spp.), forbs (e.g., *Symphyotrichum chilense*)	Subsurface perennially saturated from groundwater; temporary to seasonal surface flooding	Poorly drained clay soils
Seasonal Lake	See "Wet Meadow" above	See "Wet Meadow" above (slightly longer duration flooding than surrounding areas)	Poorly drained clay soils
Vernal Pool Complex	Dominated by annual plants (including a number of vernal pool endemics) adapted to varying levels/duration of inundation (e.g., *Lasthenia* spp., *Downingia* spp., *Plagiobothrys* spp., *Navarretia* spp.)	Temporarily to seasonally inundated from precipitation	Poorly drained clay and clay loam soils with subsurface claypan
Oak Savanna/Vernal Pool Complex	See "Vernal Pool Complex" above; interspersed with oaks (*Quercus* spp.)	Vernal pools/swales temporarily to seasonally inundated from precipitation	Poorly to well drained clay and loam soils; mounded topography; subsurface claypan
Valley Grassland	Grasses (e.g., *Elymus triticoides*) and forbs (e.g., *Madia* spp., *Hemizonia congesta* subsp. *lutescens*)	Infrequent flooding	Well drained soils
Oak Savanna	Valley oak (*Quercus lobata*) with ~10-25% canopy cover, grasses	Infrequent flooding	Deep, well drained alluvial soils
Oak Woodland	Valley oak (*Quercus lobata*) with ~25-60% canopy cover, grasses	Infrequent flooding	Deep, well drained alluvial soils
Mixed Conifer Forest	Douglas fir (*Pseudotsuga menziesii*), Coast redwood (*Sequoia sempervirens*)	Infrequent flooding	Well drained soils

Results

OVERVIEW

The Laguna de Santa Rosa and surrounding areas were historically characterized by a diversity of aquatic and wetland habitat types with a complex spatial distribution (Fig. 4-1, Table 4-4). A series of perennial freshwater lakes or ponds occupied the wettest portions of the Laguna, and were spaced at intervals along the main course of the waterway. Mixed riparian forests, comprised of oaks, willows, ash, and other species, bordered many of these lakes as well as portions of the mainstem Laguna channel and tributary channels. Perennial wetland types, such as valley freshwater marsh and willow forested wetland, occupied areas where high groundwater maintained year-round surface water or saturated soils; large expanses of these perennial wetland types occurred alongside the Laguna mainstem downstream of present-day Occidental Road. Seasonal wetland types, such as wet meadow, vernal pool complexes, and seasonal lakes, occurred in slightly higher-elevation areas characterized by poor drainage and seasonal flooding. Portions of the study area on the borders of the Laguna also supported terrestrial or upland vegetation types, such as valley grassland, oak savanna, oak woodland, and mixed conifer forest.

Table 4-4. Extent of habitat types in the Laguna study area historically.

Habitat Type	Area (ha)	Area (acres)	Percent Study Area
Oak Savanna/Vernal Pool Complex	1050	~2600	31%
Wet Meadow	890	~2190	26%
Valley Freshwater Marsh	320	~780	9%
Willow Forested Wetland	310	~770	9%
Valley Grassland	280	~700	8%
Mixed Riparian Forest	220	~550	7%
Oak Savanna	140	~350	4%
Vernal Pool Complex	110	~260	3%
Perennial Freshwater Lake/Pond	50	~130	2%
Oak Woodland	4	~10	<1%
Seasonal Lake	2	~5	<1%
Mixed Conifer Forest	0.9	~2	<1%

Figure 4-1 (right). Historical habitat types and channels within the Laguna de Santa Rosa study area representing average dry-season conditions, ca. 1850. Modern towns and road network are shown for reference.

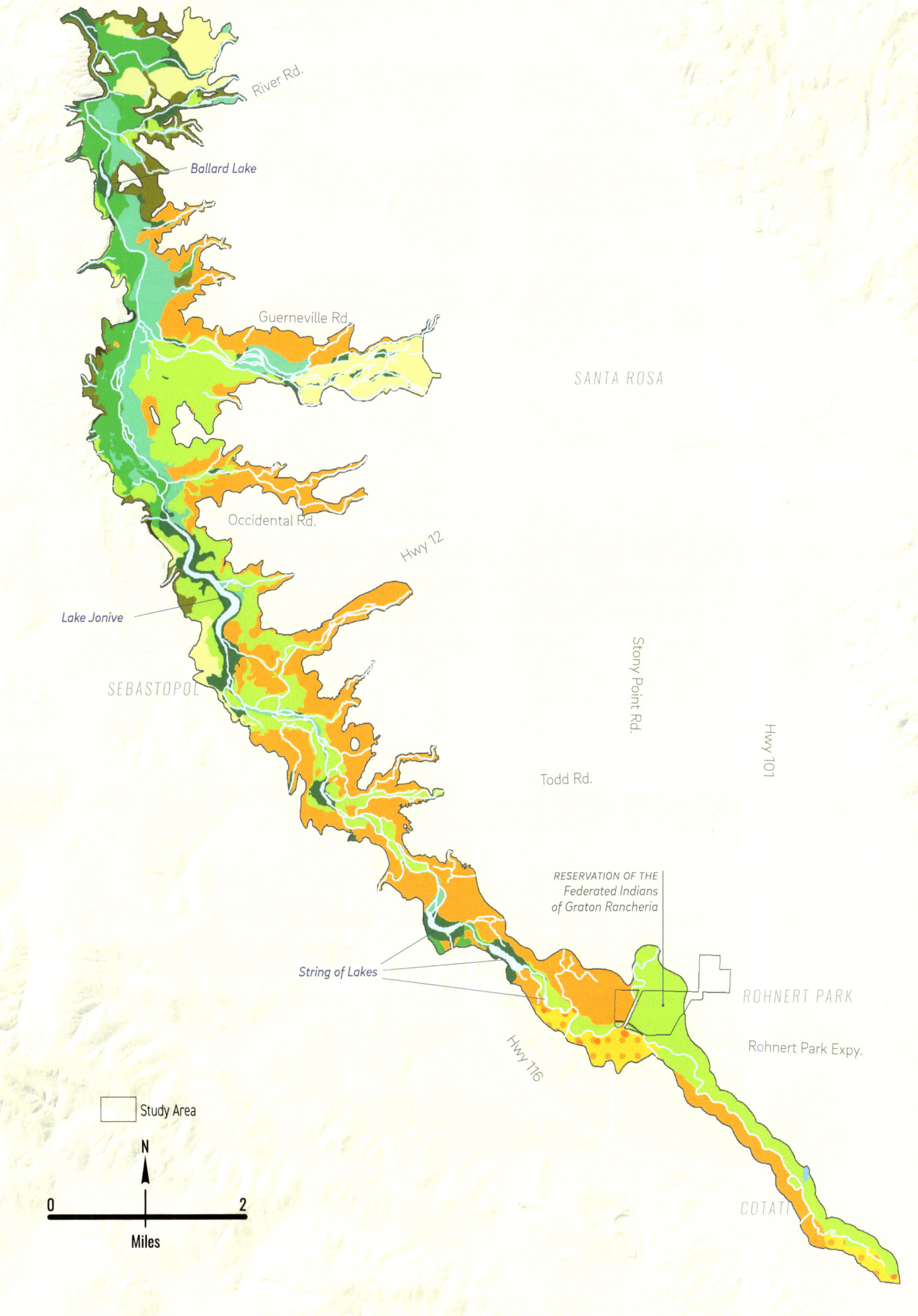
River Rd.
Ballard Lake
Guerneville Rd.
SANTA ROSA
Occidental Rd.
Hwy 12
Lake Jonive
SEBASTOPOL
Stony Point Rd.
Hwy 101
Todd Rd.
RESERVATION OF THE
Federated Indians
of Graton Rancheria
String of Lakes
ROHNERT PARK
Rohnert Park Expy.
Hwy 116
Study Area
N
0
2
Miles
COTATI

From its headwaters near the Petaluma River watershed divide (near present-day Railroad Ave), the mainstem channel of the Laguna flowed in a generally south-to-north direction through the study area as it does today. The southern portion of the mainstem was characterized by a meandering, single-threaded channel, but as the Laguna entered the more perennial wetland reaches further downstream, the channel pattern became more complex, with multiple branching channels and shallow sloughs. Numerous tributaries flowed into the Laguna wetland complex from both the Santa Rosa Plain to the east and the hills of the Wilson Grove Formation to the west.

The Laguna was characterized by pronounced seasonal variability in flow and extent of inundation. During wet-season floods, shallow open water likely covered much of the mapped wetland area: accounts of 19th and early 20th century floods, for instance, describe flooding between half a mile and several miles wide (*Daily Alta California* 1866, Menefree 1873, *Daily Courier and Petaluma Imprint* 1895 in Cummings 2006, Cardwell 1958). The Laguna was also affected by flooding on the Russian River downstream: floodwaters (and fine sediment) from the Russian River often backed up into the Laguna watershed during high flows. By late summer, however, the extent of open water would have been confined to perennial lakes and ponds and perennial portions of the channel (Millington 1865, Baumgarten et al. 2017).

The Laguna was a highly productive ecosystem that supported a variety of resident and migratory wildlife. The abundance of wildlife made the Laguna a lucrative site for early trappers and, later, a popular destination for hunters and sport fishers. A deputy surveyor for the General Land Office described the Laguna as "abound[ing] with speckled trout" (Millington 1865), while a local newspaper advertised that "persons fond of fishing, can find an abundance of trout in all our streams" (Meyer 1877). Each winter, hundreds of thousands of migratory waterfowl overwintered in the Laguna's lakes and wetlands (Laguna TAC 1989). Early observers reported that "wild fowl were plentiful...in the marshes" (Marryat 1855), while local news outlets described the region as "alive with ducks" and attested to the "excellent hunting on the Laguna" *(Petaluma Weekly Argus* 1881, *Petaluma Courier* 1883). Similarly, each summer, the Laguna provided nesting habitat for various neotropical migrant bird species such as the Western Yellow-billed Cuckoo *(Coccyzus americanus occidentalis)*, Wilson's Warbler (*Cardellina pusilla*), and Swainson's Thrush (*Catharus ustulatus*). Grizzly bears and large herbivores such as deer, tule elk, and pronghorn antelope frequented the Santa Rosa Plain (Marryat 1855, *Sonoma County Democrat* 1861).

Yellow-billed Cuckoo. **Photo:** Melissa McMasters, Creative Commons.

Tule elk, San Luis Wildlife Reserve. **Photo:** Steve Martarano, USFWS.

FRESHWATER LAKES AND RIPARIAN FORESTS

Historically, perennial freshwater lakes and ponds covered only 2% (50 ha, ~130 acres) of the Laguna, but nonetheless formed some of the landscape's most distinctive features. They were prominently illustrated in some of the earliest depictions (Fig. 4-2), and were noted by early explorers in the area. For example, writing of his travels across the Santa Rosa Plain in September of 1810, Gabriel Moraga described "a lagoon and a stream with many pools of retained water" (Moraga 1810).

The largest of the Laguna's perennial lakes, Lake Jonive, extended approximately 3 km (2 mi) from present-day Highway 12 to Occidental Road (Fig. 4-3). Early sources describe Lake Jonive as approximately 40-80 m (150-250 ft) wide (Millington 1865; *The Sebastopol Times* 1903b), and measurements taken ca. 1913 found it to have a maximum depth of 7 m (23.5 ft; Holway 1913). The lake was bordered by a mixed riparian forest composed of willows and a "heavy growth of oak" (*The Sebastopol Times* 1903a, 1903b). The northernmost of the Laguna's lakes, Ballard Lake, was located east of Vine Hill about 2 km (1.25 mi) north of Guerneville Road. Ballard Lake measured approximately 600 m (2,000 ft) in length, 70 m (250 ft) in width, and was reported to have a maximum depth of 7.6 m (25 ft; *Petaluma Weekly Argus* 1885, Holway 1913).

In the southern part of the study area, a string of 7-8 unnamed ponds, ranging in size from approximately 0.5-8 ha (~1-20 acres), dotted the Laguna channel, roughly between present-day Llano Road and Stony Point Road (Tracy 1859a; Dyer 1861). The ponds supported a dense riparian forest dominated by willows (Fig. 4-4), which served as breeding habitat for the Western Yellow-billed Cuckoo. In a 1911 paper discussing the nesting habits of cuckoos in this area, the biologist Alfred Shelton described the Laguna's perennial ponds and riparian forests as follows:

> In summer [the Laguna] is marked by a chain of long, rather narrow ponds, many of which are deep. The banks, and much of the intervening space between these ponds, are covered with a thick growth of willow, small ash and scrub oak, while the whole is tangled together with an undergrowth of poison-oak, wild blackberry and various creepers, forming, as it were, an impenetrable jungle, hanging far out over the water. –Shelton 1911

The lakes of the Laguna likely represented sag ponds which

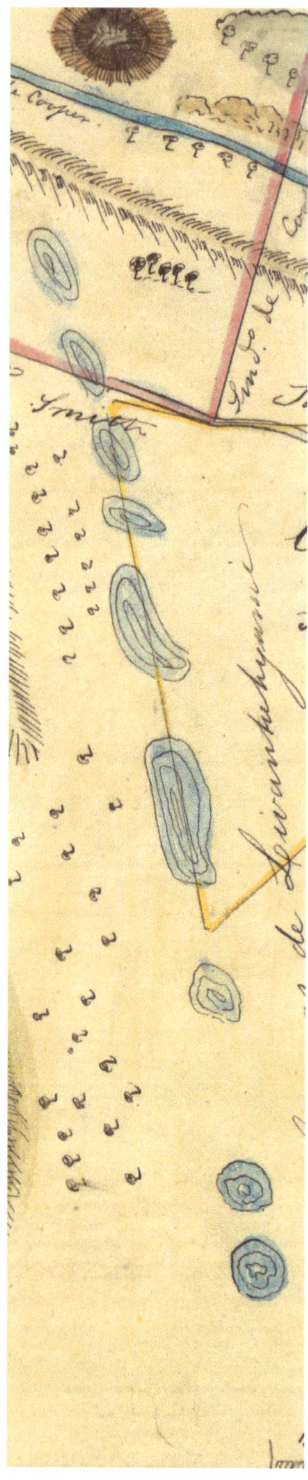

Figure 4-2. A ca. 1840 "diseño," or sketch, of Rancho Llano de Santa Rosa, showing a string of perennial lakes and ponds along the course of the Laguna de Santa Rosa. (Land Case Map B-128, courtesy of The Bancroft Library)

Figure 4-3. "Man in rowboat on Lake Jonive," 1900. (courtesy of Sonoma Heritage Collection-Sonoma County Library)

formed along the Sebastopol fault, a series of short faults which roughly parallel the course of the Laguna on the western side of the Santa Rosa Plain (Nishikawa et al. 2013). Unlike the predominantly warm, shallow open water bodies that exist in the Laguna today, these were cold, deep lakes that supported cold water fishes: the *Petaluma Weekly Argus* (1885), for instance, noted that "the water is very deep and always cold" in "Gray's Lake" (an early name for Ballard Lake). Early sources suggest that these lakes provided habitat for "salmon-trout" (steelhead; *The Sebastopol Times* 1903a), "terrapins" (pond turtles; *Healdsburg Enterprise* 1890), and many other species of fish and wildlife.

Though mostly unvegetated, portions of lakes and ponds supported submersed aquatic plants such as shining pondweed (*Potamogeton illinoensis*; Baker 1899, 1900), which is typical of deep, cold, oligotrophic lakes (Best et al. 1996, P. Baye pers. comm.); small pondweed (*P. pusillus*) was also reported (Baker 1889). The lakes were fringed in many areas by marsh vegetation that provided rich wildlife habitat. Early collectors, for instance, noted a number of nesting birds in the marshes along the shoreline, including Virginia Rail (*Rallus limicola;* "nest... in brush of wire grass... along shore of Laguna"), Common Yellowthroat (*Geothlypis trichas*; "nest... in tuft of marsh grass in shallow water of Laguna"), Marsh Wren (*Cistothorus palustris*; "nest of marsh grasses, tule and cattail bark... in tules in deep water along Laguna"), Swainson's Thrush (*Catharus ustulatus*; "nest... [in] small willow in thicket along Laguna"), Wilson's Warbler (*Cardellina pusilla*; "in wild blackberry vines... at edge of Laguna"), and Red-winged Blackbird (*Agelaius phoeniceus*; "Santa Rosa Lagoon... nest... in center of some weed stalks, built of tule grass") (van Fleet 1917; Wells 1920a, 1920b, 1923a, 1923b, 1926).

Figure 4-4. One of the small perennial freshwater lakes in the Laguna between Sebastopol and Cotati. Titled "Haunts of the California Cuckoo, in Sonoma County." (Shelton 1911)

PERENNIAL WETLANDS

Perennial wetlands, including willow forested wetland and valley freshwater marsh, occupied large areas of the Laguna floodplain, particularly to the north of Lake Jonive; small patches of valley freshwater marsh also bordered the Laguna channel upstream to Llano Road. Valley freshwater marshes are persistent emergent wetlands typically dominated by tules and bulrushes (*Bolboschoenus and Schoenoplectus* spp.), cattails (*Typha* spp.), sedges (*Carex* spp.), and rushes (*Juncus* spp.). These wetlands are seasonally to semi-permanently flooded; their soils generally have a high organic content and are usually saturated. Willow forested wetlands, as the name suggests, are dominated by willows (*Salix* spp.), but also include other tree species such as Oregon ash (*Fraxinus latifolia*) and herbaceous plants such as tule (*Schoenoplectus* spp.). These wetlands experience shallow flooding on a seasonal to semi-permanent basis, and typically occur in large stands rather than as thin corridors of riparian vegetation. Valley freshwater marsh and willow forested wetland each accounted for 9% (320 and 310 ha, ~780 and ~770 acres, respectively) of the Laguna study area historically.

Early observers and surveyors described portions of these perennial wetlands in a variety of terms, referring to "willow thickets," "swamp and tule," "marsh," "wet land of the lagoon," "bottom land," and "willow and ash timber interspersed with tule" (Whitacre 1853, Gray 1857, Tracy 1859a, Millington 1865). While there was a substantial amount of heterogeneity within and intermixing between these wetland types, in general, willow forested wetland was more common on the western side of the Laguna, while valley freshwater marsh was more common in the east (Fig. 4-5). This pattern may have been due in part to the fact that the western side of the Laguna was bordered by the steeper hills of the Wilson Grove Formation, which would have supplied both coarser sediment deposits and less freshwater input than the tributaries draining from the Santa Rosa Plain on the east. Identifying a precise historical boundary between valley freshwater marsh and willow forested wetland was difficult in many areas, which is reflected in the shape and location certainty levels for these features in the GIS mapping.

Early botanical records provide insight into the plant diversity found in these perennial wetlands (see Appendix B). Marsh plants reported from the Laguna include bur reeds (*Sparganium* spp.), hemlock waterparsnip (*Sium suave*), western water hemlock (*Cicuta douglasii*), broad leaf arrowhead (*Sagittaria latifolia*), water parsley (*Oenanthe sarmentosa*), northern water plantain (*Alisma triviale*), and many others (Best et al. 1996, records from Consortium of California Herbaria). Torrey et al. (1857) reported numerous sedge (*Carex*) species from "swamps" or "wet places" along Mark West and Santa Rosa creeks. River bulrush (*Bolboschoenus fluviatilis*) was reported to form "extensive stands" in the Laguna between Sebastopol and Trenton (Rubtzoff 1964).

Several factors helped sustain the large areas of perennial wetlands downstream of Lake Jonive. In general, these were areas characterized by poorly drained, clay soils, which tend to retain water and support emergent marsh vegetation. The geologic constriction of the Laguna channel near Trenton, along with alluvial fan deposits and sediment plugs formed by Mark West Creek, Santa Rosa Creek, and other tributaries, resulted in flow accumulation and ponding upstream (PWA 2004a); springs and high groundwater levels provided a key source of water during the dry season (Davis 1887, Cardwell 1958). Combined with the gradual slope of the Laguna, these processes caused water to move very slowly through the system and spread out over the large wetland complexes. Beaver dams also likely helped shape the historical Laguna landscape, altering the flow of water and contributing to the formation and maintenance of ponds, marshes, and other wetland habitats. Beaver were abundant in the Laguna historically: Mariano Vallejo, for instance, wrote in 1833 of "great tulare lakes teeming with beaver" (Vallejo et al. 2000), while Jose Figueroa (1834) found the Laguna to have "many beavers."

In most cases, the Laguna did not maintain a well-defined channel through these perennial wetland complexes, but rather flowed through a shallow network of anastomosing sloughs or channels (a configuration known as "stage zero" morphology; Cluer and Thorne 2014). For example, Sarepta Ann Turner Ross, an early pioneer whose family came west in a wagon train and moved to Sebastopol in 1854, recalled that near the present-day Occidental Road crossing, "there were no channels, and the water spread out for quite a long distance" (Ross 1914). Just south of present-day Highway 12, GLO surveyor Thomas Whitacre described the Laguna mainstem as a "swale" just over 30 feet wide, suggesting that it was quite shallow (Whitacre 1853).

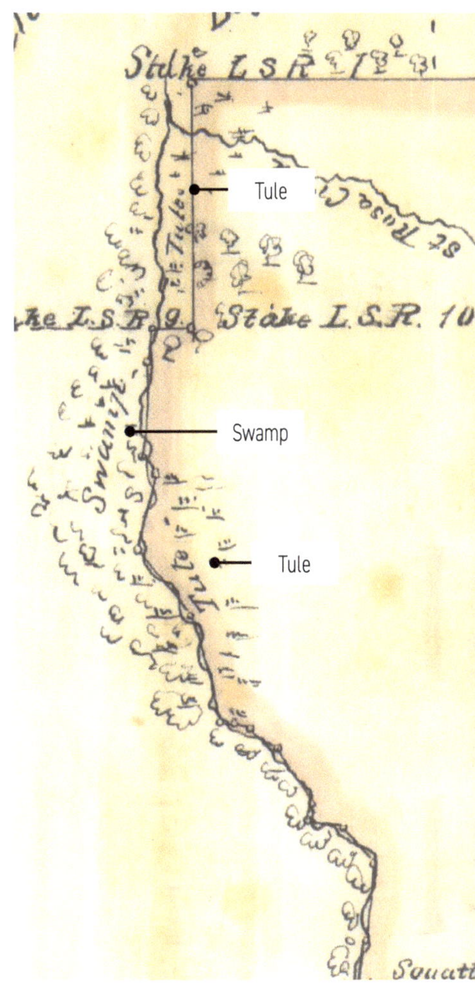

Figure 4-5. This 1859 survey plat of Rancho Llano de Santa Rosa shows valley freshwater marsh, depicted with a marsh symbol and labeled "tule," on the east side of the Laguna, and willow forested wetland, depicted with a tree symbol and labeled as a "swamp," on the west side. (Tracy 1859b, courtesy of The Bancroft Library, UC Berkeley)

SEASONAL WETLANDS

Seasonal wetlands, including wet meadows, vernal pool complexes, and seasonal lakes, occupied drier portions of the Laguna floodplain adjacent to perennial wetlands and riparian forests. Because they were dry for much of the year, seasonal wetlands were often overlooked in early maps and textual accounts, and thus there is much less evidence for these features in the historical record than there is for more perennial wetland types. However, the historical distribution of seasonal wetlands can be reconstructed with reasonable accuracy from a combination of historical aerial photographs, early soil surveys, the LiDAR-derived DEM, and other sources.

Extensive areas of wet meadow bordered the Laguna in the central and southern portions of the study area, representing 26% (890 ha, ~2190 acres) of the total mapped area. This seasonal wetland type occurred in areas with poorly drained, clay-rich soils and perennially high groundwater levels, and during the wet season, experienced temporary or seasonal flooding. Wet meadows supported an herbaceous plant community dominated by perennial grasses (e.g., *Elymus triticoides*), rushes (*Juncus* spp.), spike-rushes (*Eleocharis* spp.), sedges (*Carex* spp.), and forbs such as pacfic aster (*Symphyotrichum chilense*), western goldenrod (*Euthamia occidentalis*), mugwort (*Artemisia douglasiana*), sneezeweed (*Helenium* spp.), wholeleaf saxifrage (*Micranthes integrifolia*), and wild mint (*Mentha arvensis*) (Torrey et al. 1857, Rubtzoff 1964, Best et al. 1996, records from Consortium of California Herbaria). The largest tracts of wet meadow occurred south of the Laguna's confluence with Santa Rosa Creek, including a large, contiguous swath of the Cotati Plain that extended beyond the study area. Historical wet meadows largely coincided with Dublin adobe and Dublin clay adobe soils, which are notable for their poor drainage (Fig. 4-6; Holmes and Nelson 1914, 1915; Watson et al. 1915, 1917).

A series of seasonal lakes occurred within the large expanse of wet meadow on the Cotati Plain; one of these is included within the study area (Fig. 4-7; Martin 1859, Unknown ca. 1870). These lakes likely occupied the wettest, lowest-elevation areas of the wet meadow complex, and as such would have retained water slightly longer than surrounding areas as seasonal floodwaters evaporated. Though these lakes may have functioned similarly to vernal pools in the duration and timing of seasonal flooding, they appear much larger and more well-defined than features in the vernal pool complexes to the north and do not appear to have had the subsurface hardpan characteristic of vernal pools.

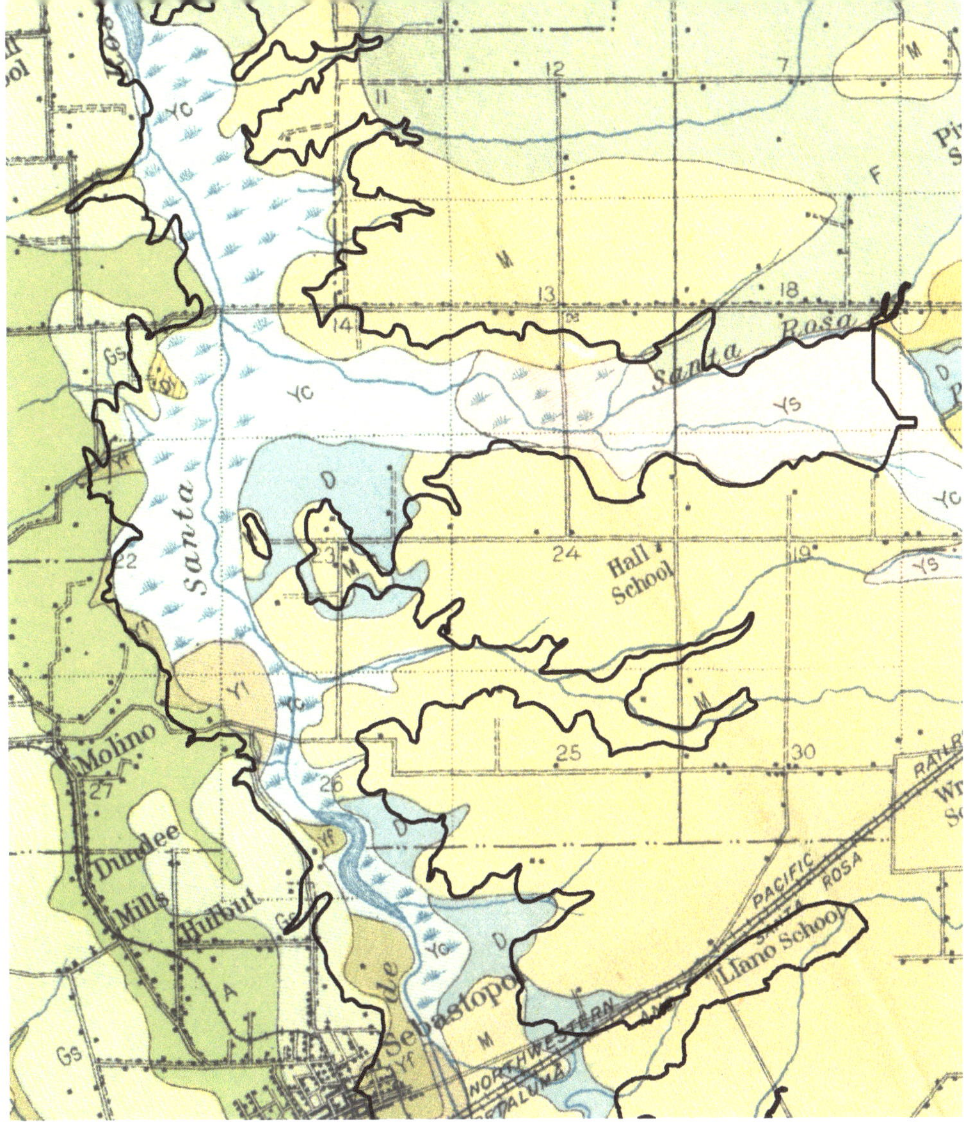

Figure 4-6. (above) 1915 soil map showing the general distribution of perennial freshwater marsh and forested wetland as indicated by the tufted "marsh" symbol, primarily in areas of Yolo silty clay loam (Yc). Wet meadows typically occurred in areas of Dublin adobe (D) or Dublin clay adobe soils (not shown), while vernal pool complex occurred in areas of Madera loam (M) or Fresno loam (not shown). Thick black boundary shows a portion of the study area. (Watson et al. 1915, courtesy of University of Alabama)

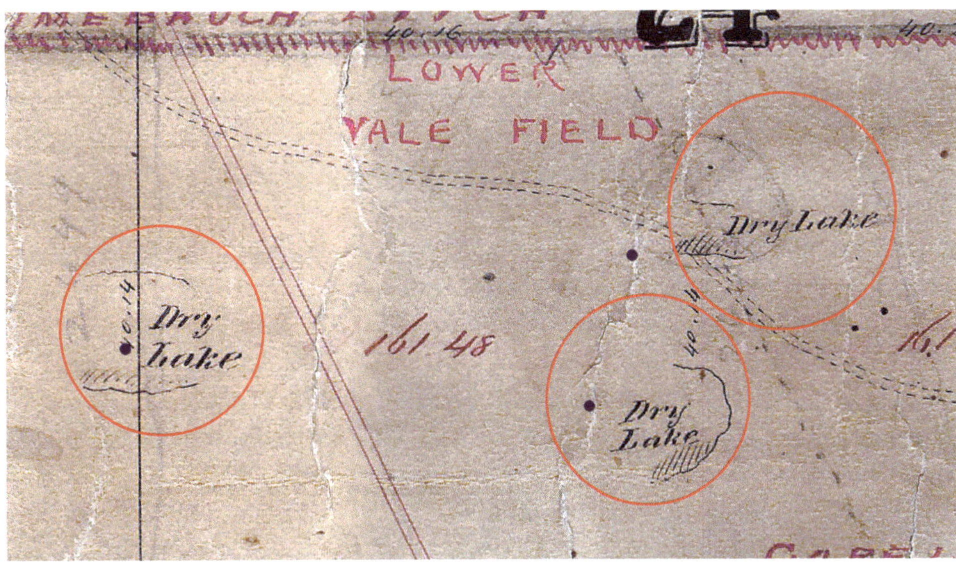

Figure 4-7. 19th century map of "Cotate Rancho" showing "dry lakes" within the surrounding wet meadow complex on the Cotati Plain. (Unknown ca. 1870, courtesy of Curtis and Associates, Inc.)

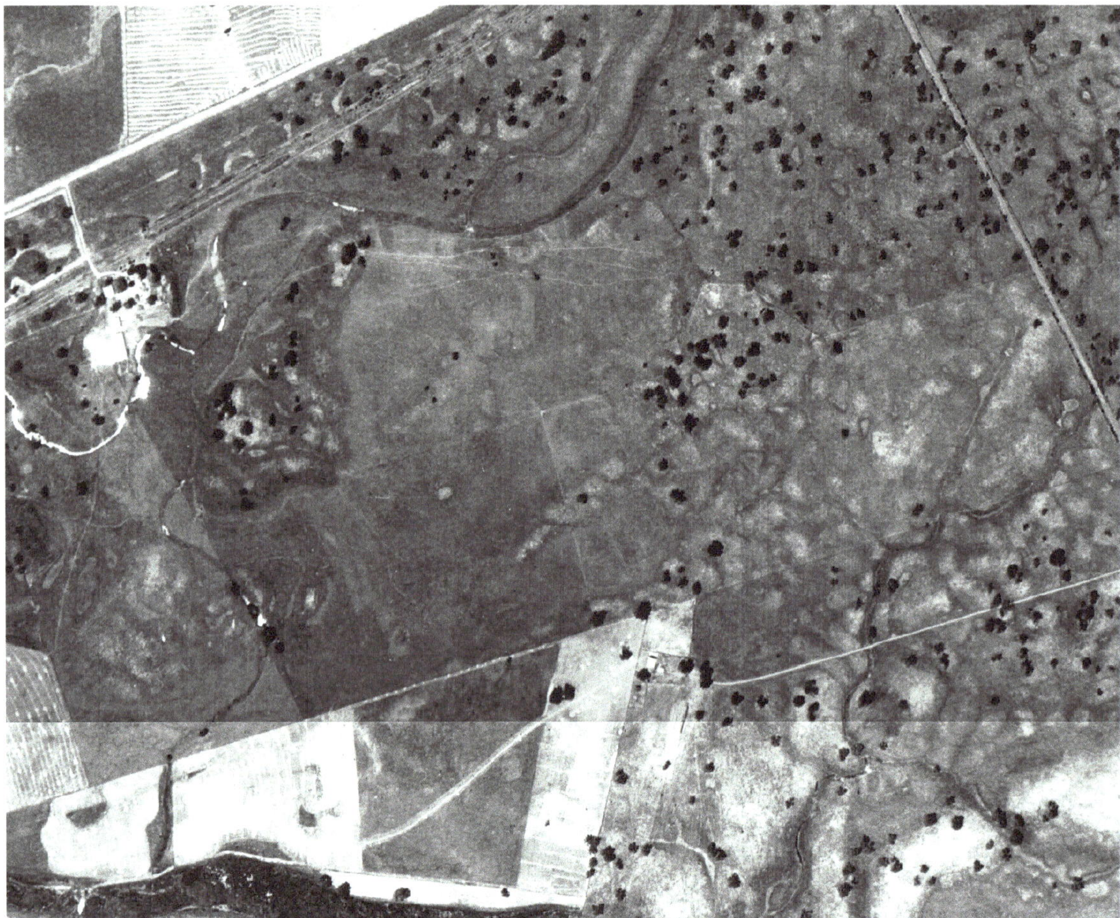

Figure 4-8. 1942 aerial photo showing oak savanna/vernal pool complex on the Santa Rosa Plain.

Vernal pool complexes, consisting of numerous interconnected pools and swales, and often intermixed with oak savanna, were among the most extensive habitat types documented in the study area, occupying approximately 1,156 ha (2,857 ac) historically (34% of the study area; Fig. 4-8). Vernal pools occur in areas with mounded topography and an impermeable subsoil layer or "claypan"; during the wet season, standing water accumulates in the poorly drained depressions and persists for varying amounts of time depending on pool depth. Unlike wet meadows, vernal pool soils typically dessicate during the dry season, and thus the plant community is dominated by annual or dry-dormant species (Barbour et al. 2007). Vernal pools on the Santa Rosa Plain supported a diverse and specialized flora, including species such as flatface calicoflower (*Downingia pulchella*), tricolor monkeyflower (*Diplacus tricolor*), common meadowfoam (*Limnanthes douglasii*), Lobb's aquatic buttercup (*Ranunculus lobbii*; CNPS Rare Plant Rank 4.2), and Baker's navarretia (*Navarretia leucocephala* ssp. *bakeri*; CNPS Rare Plant

Rank 1B.1), as well as four federally and state-endangered plants (see page 5; Torrey et al. 1857, Robbins 1937, Rubtzoff 1966, USFWS 2005, records from Consortium of California Herbaria). Vernal pools and other wetland habitats in the Laguna also supported California tiger salamander (sometimes referred to as "water dogs" in early sources; e.g., *Sonoma Democrat* 1879), which is now endangered in the region.

The distribution of historical vernal pool complexes was identified primarily from early soil surveys (Holmes and Nelson 1914, Watson et al. 1915, see Fig. 4-6) and aerial photos (Fig. 4-8). Watson et al. (1917), for instance, provides this description of a characteristic vernal pool soil: "The surface [of Madera loam] is usually uneven, as the result of the occurrence of numerous small mounds and intervening depressions. These retain water during the rainy season, owing to the impervious subsoil and hardpan... The original growth on this soil consisted of grasses and scattered valley oaks."

In the northern part of the study area, oak savanna/vernal pool complexes typically occurred on higher-elevation parts of the floodplain bordering the Laguna to the east. Soil surveys, aerial photos, and other sources suggest that these habitats were contiguous with a much larger area of oak savanna/vernal pool complex that occupied much of the Santa Rosa Plain historically. In the southern part of the study area near Cotati, however, vernal pool complex and oak savanna/vernal pool complex instead dominated the western side of the Laguna: the Laguna channel marked the approximate divide between oak savanna/vernal pool complex on the west and wet meadow on the east.

TRIBUTARIES AND RIPARIAN CORRIDORS

Numerous tributaries drained into the Laguna from both the Santa Rosa Plain to the east and the hills to the west. In many cases, these tributaries were bordered by a combination of riparian forests and perennial and seasonal wetlands. Santa Rosa Creek was, and remains today, the largest tributary to the Laguna. It connected with the Laguna mainstem just south of present-day Guerneville Road. In its lower reaches (those included within the historical synthesis mapping), Santa Rosa Creek was characterized by an anastomosing channel morphology with multiple branching channels and side channels that flowed through a floodplain approximately 1,000 m (3,000 ft) wide (Millington 1865, US Surveyor General's Office 1865, USDA 1942). Mixed riparian forests lined many of the channel segments along Santa Rosa Creek, and were likely more extensive than the remnant patches of forest documented in the historical synthesis mapping (USDC ca. 1840, ca. 1849). A large freshwater marsh occupied much of the Santa Rosa Creek floodplain around present-day Willowside Road (Watson et al. 1915, 1917).

Mark West Creek entered the study area in the northeast, and flowed west to its confluence with the Laguna approximately 0.8 km (0.5 mi) north of present-day River Road, near Trenton (Whitacre 1853, Bowers 1866, US Surveyor General's Office 1868,

Figure 4-9. Map of Rancho San Miguel (ca. 1840) showing riparian corridor along Mark West Creek. (USDC ca. 1840, courtesy of The Bancroft Library, UC Berkeley).

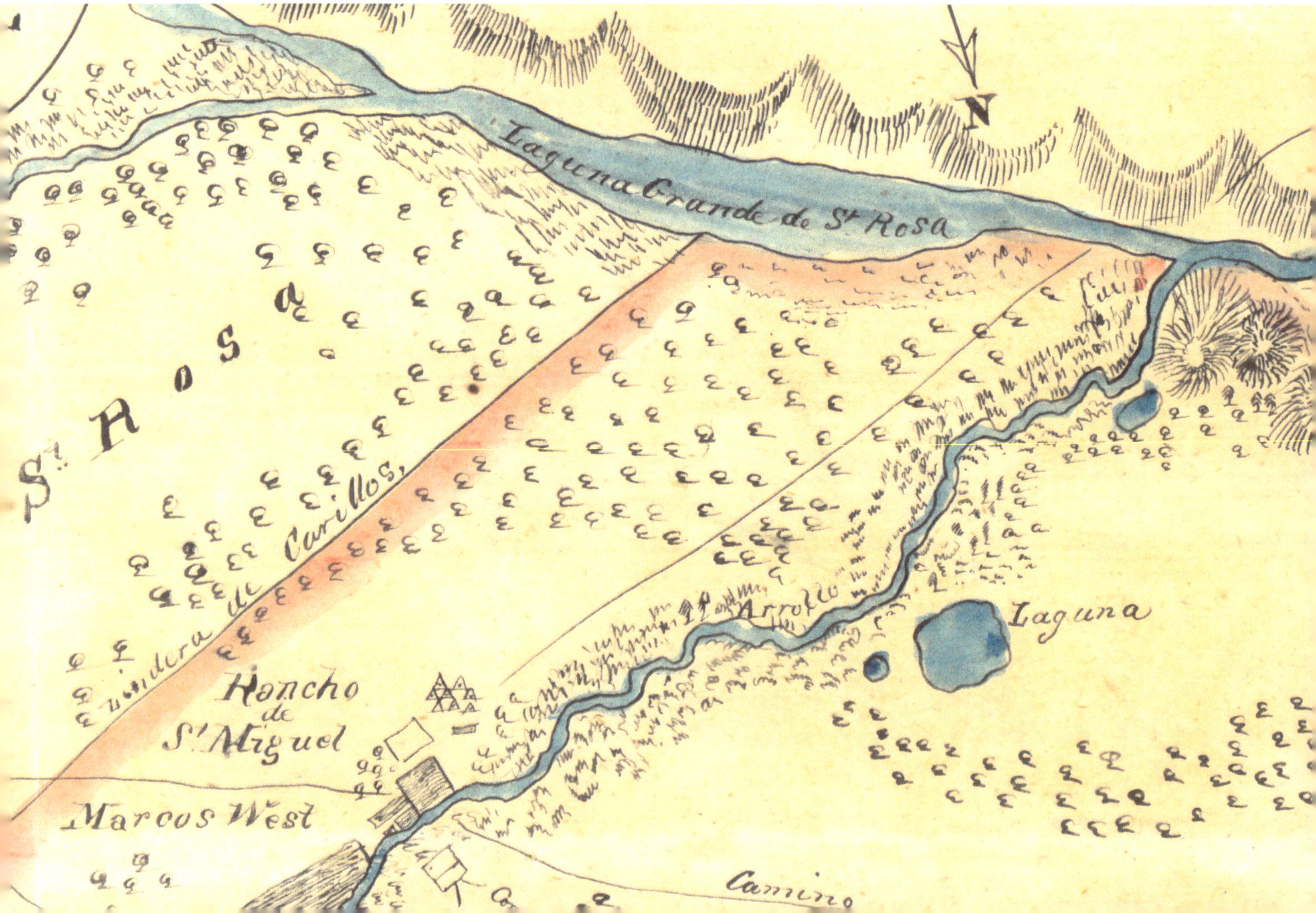

Baumgarten et al. 2014). The creek likely shifted periodically across its alluvial fan in this area, with flows sometimes moving south into the Woolsey Creek drainage. Mark West Creek supported a broad heterogeneous mixture of riparian forest and willow forested wetland (Fig. 4-9). Just upstream of the study area on Mark West Creek, GLO surveyor Seth Millington described riparian forests dominated by maple and ash (Millington 1865). Further downstream, surveyor Nicholas Gray reported the "low grounds of Mark West Creek" to be dominated by oak, with a "thick under growth of vines and brier [sic]" (Gray 1857). Near the confluence of the Laguna, the riparian forests along Mark West Creek gave way to "willow thickets" and "marsh land" (Whitacre 1853).

Several smaller tributaries south of Mark West Creek, such as Woolsey Creek and Rued Creek, supported wetland complexes comprised of small perennial freshwater lakes surrounded by valley freshwater marsh, wet meadow, and mixed riparian forests. Described as "spring branch[es]" (Davis 1887), these spring-fed creeks originated on the western side of the Santa Rosa Plain where high groundwater emerged at the base of the alluvial fan deposits, and provided sufficient flow to maintain perennial wetlands over a mile inland from the mainstem Laguna channel (Fig. 4-10).

Tributaries such as Mark West Creek provided spawning habitat for a number of native fish species, including steelhead and coho salmon (CDFG 2004, Spence et al. 2005). Early newspaper articles, for instance, reported that "salmon trout are plentiful in Mark West Creek" (*Sonoma Democrat* 1882) and that "large salmon" had also been caught (*Press Democrat* 1886). The *Sonoma Democrat* (1875) noted that "salmon trout run up these streams [on the east side of the Santa Rosa Plain] nearly to their source to spawn."

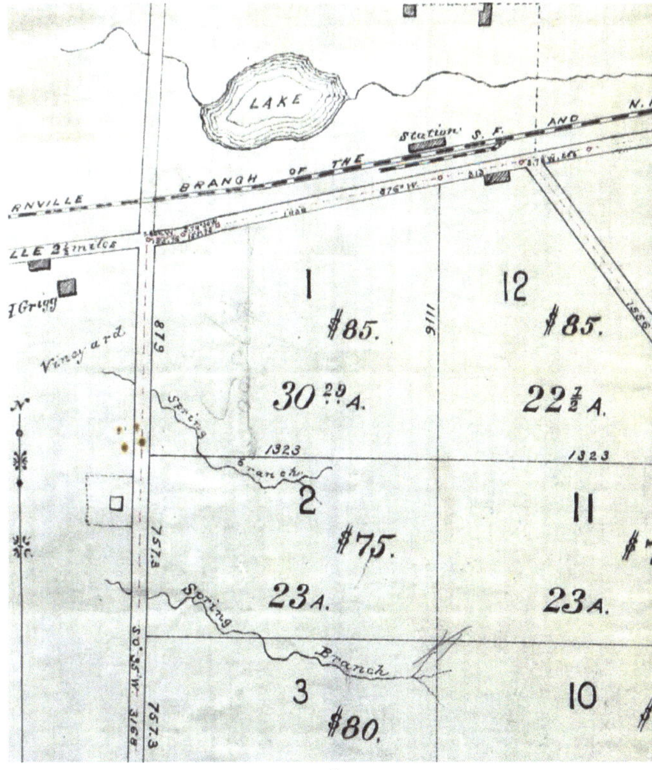

Figure 4-10. 1887 map showing lake and "spring branch[es]" along Woolsey and Rued creeks on the Santa Rosa Plain east of the Laguna. (Davis 1887, courtesy of Curtis and Associates, Inc.)

UPLAND HABITATS

Upland habitats historically occurred on the periphery of the study area, in areas that experienced less frequent flooding. Given the focus of this Vision on the Laguna's 100-year floodplain, historical land cover within the study area was dominated by wetland, aquatic, and riparian habitat types; upland habitat types, which included valley grasslands, oak savannas and woodlands, and mixed conifer forests, together accounted for approximately 13% (430 ha, ~1060 acres) of the study area.

Valley grasslands, characterized by grasses (e.g., *Elymus triticoides*) and forbs (e.g., *Madia* spp., *Hemizonia congesta* subsp. *lutescens*) and generally lacking woody shrubs or trees, were the most abundant upland habitat within the study area historically. They occupied 8% (280 ha, ~700 acres) of the study area and occurred in large patches in the northern half of the floodplain, most notably among the braided channels of Santa Rosa Creek. Oak dominated habitats, with an overstory comprised of valley oak and other oak species (*Quercus* spp.), and an understory dominated by grasses and forbs, accounted for 4% (144 ha, ~360 acres) of the study area historically. These habitats ranged from sparser oak savannas (~10 to 25% canopy cover) to denser oak woodland (~25 to 60% canopy cover). They generally occurred at higher elevations in the northern part of the study area, including many areas near Mark West Creek.

Early observers marvelled at the size of the oaks and the luxuriance of the valley grasslands on the Santa Rosa Plain and surrounding areas. Riding on horseback in the vicinity of the Santa Rosa Plain in 1833, for instance, F. P. Von Wrangell described "immense meadows, where the lushest kind of grass grew abundantly," as well as "superb oak forests, neat as an English park, [which] alternated with lush meadows" (Von Wrangell et al. 1974). Further on he described a plain "luxuriantly overgrown with fragrant herbs" with "magnificent oak groves that provide shadow to the plain here and there," and noted that "the horses almost disappeared in the tall, fragrant grass, which covers the meadow."

Small patches of mixed conifer forest occurred at high elevations along the periphery of the northern Laguna, accounting for <1% (0.9 ha, ~2 acres) of the study area. Mixed conifer forests include forests and woodlands dominated by Douglas fir (*Pseudotsuga menziesii*), coast redwood (*Sequoia sempervirens*), or other coniferous trees.

Summary

The examination of how the Laguna looked and functioned in the recent past provides the basis for understanding how the Laguna landscape has changed over the past 200 years (Chapter 5). While the term "Laguna" conjures an image of a single discrete feature, in reality the historical Laguna de Santa Rosa encompassed a diverse assemblage of intergraded habitat types, including deep lakes and ponds, perennial wetlands, seasonal wetlands, riparian forests, and upland transition zones. The diversity and complexity of habitat types that the Laguna supported, and their spatial arrangement on the landscape, were a direct result of physical processes (hydrologic, geomorphic, and tectonic), which varied both spatially and temporally, and physical gradients (elevational and edaphic). Perennial wetlands occupied lower elevation areas that received year-round inputs of surface or groundwater, while seasonal wetlands occupied poorly drained soils in slightly higher elevation areas on the periphery of the Laguna. Deep lakes and ponds formed where tectonic movement created depressions along the Sebastopol fault, while broad riparian forests dominated the margins of these aquatic habitats. The scale and diversity of the Laguna's habitats, in turn, provided resources that supported an immensely rich and abundant wildlife community, including salmonids and other native fish, vast numbers of migratory waterfowl and neotropical songbirds, and a wide range of mammals, reptiles, amphibians, and invertebrates. §

5 Landscape Change Analysis

View of Balletto Vineyards from Laguna de Santa Rosa Trail. Photo: Harminder Dhesi. CC by SA 2.0.

Introduction

The Laguna has experienced major physical and ecological changes over the past two centuries. Urban and agricultural development, draining and filling of wetlands, levee construction and channel straightening, and other land and water use modifications have altered habitat extent, distribution, and channel configuration throughout the study area and the broader watershed. This chapter examines the types, magnitude, and effects of landscape changes in the Laguna study area by analyzing changes in habitat extent, habitat configuration, and channel planform, as well as the impacts of invasive species introductions.

OVERVIEW OF LANDSCAPE CHANGE

The Laguna region is believed to have been inhabited by humans for at least the past 7,000 years (Origer and Frederickson 1980), though there is unpublished evidence indicating possible human presence far earlier (T. Origer, pers. comm.). At the time of earliest European contact, the Laguna de Santa Rosa region was home to the Southern Pomo and Coast Miwok Indians. A number of villages were located near the Laguna, which provided a productive resource for food and materials for these communities (Barrett 1908, Kroeber 1925, Origer and Fredrickson 1980, Fredrickson and Markwyn 1990). Stewart (1943), for instance, notes that the Bitakomtara tribelet on the Santa Rosa Plain obtained much of their fish from "creeks and laguna" and constructed "willow-frame houses," while the Konhomtara tribelet to the west constructed temporary tule houses along the Laguna. Native tribes in the area actively managed the landscape in a number of ways, including using fire to alter vegetation cover and enhance resource yields (Werner et al. 2003, Anderson 2005, Welch 2013, T. Origer, pers. comm.); Von Wrangell, for instance, describes witnessing a "burning thicket" near a "fairly populous Indian village" in the vicinity of the Santa Rosa Plain in September of 1833 (von Wrangell et al. 1974).

European American modification of the landscape began during the early 1800s, as settlers displaced native communities, cleared large tracts of land for ranching, and decimated wildlife populations. Hunting and trapping by the Russian-American Company, the Hudson's Bay Company, and others drastically reduced the numbers of many wildlife species, contributing to the eventual extirpation of grizzly bear, pronghorn, tule elk, beavers, and other species native to the watershed (Sloop et al. 2007). Cattle grazing, the principal economic activity on the large land grants that dominated the Santa Rosa Plain in the early to mid-19th century, likely resulted in significant soil erosion and contributed to the introduction of non-native plant species (Gregory 1911, PWA 2004b). By the late 19th century, agriculture had largely supplanted ranching as the dominant economic activity in the valley: wheat, barley, oats, grapes, and other crops were widely cultivated (Walker 1880, Watson et al. 1917). Large numbers of oaks on the Santa Rosa Plain were felled to clear land for cultivation, and portions of the Laguna floodplain were likewise drained and filled for agriculture (Taylor 1862, *Pacific Rural Press* 1880, PWA 2004a). Towns such as Santa Rosa, Cotati, and Sebastopol grew rapidly in the late 19th and early 20th centuries. Sewage from septic systems and growing urban areas was often discharged directly into the Laguna or other waterways, leading to impaired water quality and public health hazards (*Daily Alta California* 1888; Lee 1944; Cummings 2003a, 2003b). Urban growth was accompanied by efforts to control flooding by ditching, channelizing, and rerouting streams. In addition, while many of the North Bay's ecosystems are adapted to periodic fires, decades of fire suppression efforts have substantially altered fire regimes in many parts of the watershed (Community Foundation Sonoma County 2010, Safford et al. 2013).

The cumulative impacts of these varied land and water use modifications on the Laguna de Santa Rosa, and the plant and animal species it supports, have been considerable. Substantial loss and fragmentation of wetland, aquatic, and riparian habitats has occurred both directly, through draining, filling, and clearing, and indirectly through changes in hydrology and sediment dynamics. In addition, many of the remaining habitats have been degraded through elevated nutrient levels, other water quality impairments (see page 3), as well as widespread establishment of *Ludwigia hexapetala* and other invasive species. These changes have decreased the Laguna's ability to support native biodiversity and to provide desired ecosystem functions such as flood protection. A number of species, such as Western yellow-billed Cuckoo and the top predators and large herbivores mentioned above, have been extirpated from the Laguna or the watershed altogether. Many other species are still present but are rare relative to their historical abundance.

Despite landscape changes over the past several centuries, the Laguna continues to provide valuable habitat for a diversity of plants and animals, and a range of ecosystem services to surrounding communities. In addition to its importance in supporting biodiversity (see page 7), the Laguna plays a key role in providing flood storage, nutrient assimilation, recreational opportunities, scenic values, and a range of other benefits (Sloop et al. 2007, Sloop and Jones 2010).

Methods

The first step in assessing landscape change over time was to assemble a map of contemporary land cover and channel locations (Fig. 5-1 on page 70). NCARI (SFEI-ASC 2014) was used to map the boundaries of most wetland, aquatic, and riparian habitat features, as well as the configuration of the contemporary channel network, within the study area. Land cover outside of the boundaries of the NCARI mapping was derived from the Sonoma Veg Map dataset (Sonoma Veg Map 2017). Modern classifications were modified (i.e., renamed or grouped together) in order to facilitate comparison between the historical and modern mapping (Table 5-1). Minor manual adjustments were made to the modern land cover map in some areas using a combination of modern aerial imagery (NAIP 2016) and local expert knowledge. Due to limitations in the specificity of the historical data, the land cover classes used in the change analysis are necessarily broad, and do not represent the full complexity and heterogeneity of wetland types present in the Laguna.

Several adjustments to the historical mapping were made to facilitate comparison with the modern mapping. First, because Willow Forested Wetland and Mixed Riparian Forest are not consistently differentiated in modern vegetation mapping, they were combined into a single class called Forested Wetland and Riparian Forest/Scrub. Second, it was not possible to consistently distinguish Oak Savanna, Oak Woodland, Valley Grassland, Vernal Pool Complex, and Seasonal Lake in the historical and modern mapping, and thus these habitat types were combined into a single class. Mixed Conifer Forest was reclassified as Other Upland.

Five novel habitat types that were not present historically were defined in the modern landscape: Agriculture (including row crops, vineyards, and orchards; pastures and hayfields were categorized as grasslands), Developed/Disturbed, Farmed Wetland (agricultural areas that flood during the winter), Storage Pond, and Non-native Aquatic/Emergent Vegetation (areas that are dominated by *Ludwigia hexapetala*, along with pockets of native aquatic/emergent vegetation, and are often shallowly flooded).

The historical and modern mapping was analyzed to calculate changes in overall habitat extent, as well as changes to the spatial distribution and arrangement of habitats and channels on the landscape. A series of metrics related to habitat configuration were developed to quantitatively measure the magnitude of these changes, in order to evaluate potential impacts on target ecological functions, ecosystem services, and the resilience of the system. Detailed description of analysis methods are included in Appendix C.

Table 5-1. Modifications made to habitat classifications in modern mapping sources to facilitate comparison with historical mapping. Modern wetland classes follow the classification systems used by NCARI (SFEI-ASC 2014) and the Sonoma Veg Map (Sonoma Veg Map 2017). Note: A 50 foot buffer of riparian forest was manually added around the mainstem channel in the upstream portion of the study area (south of Highway 101).

Source	Original Classification	Modified Classification
NCARI	Channel open water natural	1) Lake Jonive and similar large open water features –> Perennial Freshwater Lake/Pond 2) For narrow channels, deferred to adjacent habitat classification
NCARI	Channel open water unnatural	Classification was merged with adjacent polygons
NCARI	Channel vegetated natural	Valley Freshwater Marsh
NCARI	Channel vegetated unnatural	Valley Freshwater Marsh
NCARI	Depressional open water natural	Perennial Freshwater Lake/Pond
NCARI	Depressional open water unnatural	1) Wastewater treatment ponds –> Developed/Disturbed 2) Small open water features embedded within other habitat types –> Perennial Freshwater Lake/Pond 3) Managed ponds or other small, unnatural open water features outside the core of the Laguna –> Storage Pond 4) Large polygons between Occidental Road and Guerneville Road with aquatic vegetation visible in NAIP 2016 –> Non-native Aquatic/Emergent Vegetation 5) One polygon classified as Farmed Wetland based on local expert knowledge 6) One polygon classified as Forested Wetlands and Riparian Forest/Scrub based on modern aerial imagery (NAIP 2016) 7) One polygon classified as Valley Freshwater Marsh based on modern aerial imagery (NAIP 2016) and local expert knowledge
NCARI	Depressional vegetated natural	Valley Freshwater Marsh
NCARI	Depressional vegetated unnatural	1 Default = Valley Freshwater Marsh 2) Several polygons classified as Forested Wetland and Riparian Forest/Scrub based on modern aerial imagery (NAIP 2016) 3) Several polygons north of Occidental Road classified as Wet Meadow based on local expert knowledge
NCARI	Farmed depression unnatural	1) Default = Farmed Wetland 2) Several polygons classified as Wet Meadow based on consultation of Sonoma Veg Map
NCARI	Farmed slope wetland natural	1) Default = Farmed Wetland 2) Several polygons classified as Wet Meadow based on local expert knowledge and consultation of Sonoma Veg Map
NCARI	Individual vernal pool	Oak Savanna or Woodland/Vernal Pool Complex and Valley Grassland
NCARI	Lacustrine open water unnatural	Storage Pond
NCARI	Lacustrine vegetated unnatural	Valley Freshwater Marsh
NCARI	Natural slope	Wet Meadow
NCARI	Unnatural slope	Wet Meadow
NCARI	Vernal pool complex	Oak Savanna or Woodland/Vernal Pool Complex and Valley Grassland
NCARI	Wet meadow	Wet Meadow
Sonoma Veg Map	Annual cropland	Agriculture
Sonoma Veg Map	Baccharis pilularis alliance	Other Upland
Sonoma Veg Map	Barren–sparsely vegetated	Developed/Disturbed
Sonoma Veg Map	California annual and perennial grassland	Oak Savanna or Woodland/Vernal Pool Complex and Valley Grassland
Sonoma Veg Map	Developed	Developed/Disturbed
Sonoma Veg Map	Dry stock pond	Storage Pond
Sonoma Veg Map	*Eucalyptus* (*globulus, camaldulensis*) semi-natural alliance	Other Upland

Table 5-1. Continued.

Source	Original Classification	Modified Classification
Sonoma Veg Map	Forest sliver	1) Default = Forested Wetland and Riparian Forest/Scrub 2) If adjacent to Developed/Disturbed, merge with that classification
Sonoma Veg Map	Forested slope	Forested Wetland and Riparian Forest/Scrub
Sonoma Veg Map	Intensively managed hayfield	Oak Savanna or Woodland/Vernal Pool Complex and Valley Grassland
Sonoma Veg Map	Irrigated pasture	Oak Savanna or Woodland/Vernal Pool Complex and Valley Grassland
Sonoma Veg Map	Major roads	Developed/Disturbed
Sonoma Veg Map	Non-native forest–woodland	Other Upland
Sonoma Veg Map	Non-native shrub	Other Upland
Sonoma Veg Map	Nursurey or ornamental horticultural area	Agriculture
Sonoma Veg Map	Orchard or grove	Agriculture
Sonoma Veg Map	Perennial agriculture	Agriculture
Sonoma Veg Map	*Populus fremontii* alliance	Forested Wetland and Riparian Forest/Scrub
Sonoma Veg Map	*Pseudotsuga menziesii* alliance	Other Upland
Sonoma Veg Map	*Quercus* (*agrifolia, douglasii, garryana, kelloggii, lobata, wislizenii*) alliance	Oak Savanna or Woodland/Vernal Pool Complex and Valley Grassland
Sonoma Veg Map	*Quercus agrifolia* alliance	Oak Savanna or Woodland/Vernal Pool Complex and Valley Grassland
Sonoma Veg Map	*Quercus garryana* alliance	Oak Savanna or Woodland/Vernal Pool Complex and Valley Grassland
Sonoma Veg Map	*Quercus lobata* alliance	Oak Savanna or Woodland/Vernal Pool Complex and Valley Grassland
Sonoma Veg Map	Riparian–forested slope	Forested Wetland and Riparian Forest/Scrub
Sonoma Veg Map	Rubus armenicus alliance	Forested Wetland and Riparian Forest/Scrub
Sonoma Veg Map	*Sequoia sempervirens* alliance	Other Upland
Sonoma Veg Map	Southwestern north american riparian evergreen and deciduous	Forested Wetland and Riparian Forest/Scrub
Sonoma Veg Map	Southwestern north american riparian/wash scrub group	Forested Wetland and Riparian Forest/Scrub
Sonoma Veg Map	*Umbellularia californica* alliance	If adjacent to channel and riparian forest, then Forested Wetland and Riparian Forest/Scrub; if not, Mixed Conifer Forest
Sonoma Veg Map	Urban window	Developed/Disturbed
Sonoma Veg Map	Vancourverian riparian deciduous forest group	Forested Wetland and Riparian Forest/Scrub
Sonoma Veg Map	Vineyard	Agriculture
Sonoma Veg Map	Vineyard repalnt	Agriculture
Sonoma Veg Map	Water	1) Small open water features embedded within other habitat types –> Perennial Freshwater Lake/Pond 2) Managed ponds or other small, unnatural open water features outside the core of the Laguna –> Storage Pond 3) For narrow slivers, deferred to adjacent habitat classification
Sonoma Veg Map	Western north american vernal pool macrogroup	Oak Savanna or Woodland/Vernal Pool Complex and Valley Grassland
Sonoma Veg Map	Western north american freshwater aquatic vegetation macrogroup	1) Large open water features –> Perennial Freshwater Lake/Pond 2) For narrow slivers, deferred to adjacent habitat classification
Sonoma Veg Map	Western north american freshwater marsh macrogroup	1) Default = Valley Freshwater Marsh 2) For narrow slivers, deferred to adjacent habitat classification

Changes in Habitat Extent

Over the past two centuries, the distribution and extent of land cover within the study area has changed substantially (Fig. 5-1). Overall, the area occupied by wetland, riparian, and aquatic habitat types that were present historically (i.e., excluding novel habitat types) has declined by 63%. Habitat loss includes an 81% loss of Valley Freshwater Marsh, 74% loss of Wet Meadow, 37% loss of Forested Wetland and Riparian Forest/Scrub, and 39% loss of Perennial Freshwater Lakes/Ponds (Fig. 5-2). Novel land cover types, including Developed/ Disturbed, Agriculture, Storage Pond, Non-native Aquatic/Emergent Vegetation, and Farmed Wetland, today occupy 32% of the study area (see Fig. 5-2).

Notable categories of land cover change include conversion of Valley Freshwater Marsh to Farmed Wetland (81 ha, ~200 acres) and Agriculture (65 ha, ~160 acres); conversion of Wet Meadow to Oak savanna or Woodland/Vernal Pool Complex/Valley Grassland (433 ha, ~1070 acres) and Developed/Disturbed land (156 ha, ~385 acres); and conversion of Forested Wetland and Riparian Forest/Scrub to Oak Savanna or Woodland/Vernal Pool Complex/Valley Grassland (113 ha, ~279 acres) and Agriculture (98 ha, ~240 acres; Fig. 5-2 and Fig. 5-3). In general, the observed land cover changes represent conversion from wetter to drier habitat types; this trend is likely due to a combination of direct filling of wetlands, channel modifications (e.g., channelization and levee construction) that have increased drainage efficiency and reduced channel-floodplain connectivity (see page 86), and declines in groundwater levels that occurred during the late 20th century (see page 27; Nishikawa et al. 2013).

Most of the perennial lakes and ponds that existed within the Laguna historically have disappeared as a result of channelization/drainage and sediment accumulation. The broad riparian forests that existed around these waterbodies have likewise been eliminated or greatly narrowed (see page 78). Ballard Lake (see page 50) largely filled with sediment delivered by Mark West Creek during a series of floods in the 1940s (at which time Mark West Creek drained directly into Ballard Lake; Baumgarten et al. 2014); the shallow pond that remained was then drained by local landowners to alleviate flooding and mosquitos (Denner 2002). Lake Jonive has decreased in size by approximately 50% (from ~27 ha to ~14 ha/~67 acres to ~35 acres), and is shallower today than it was during the mid-19th century (Butkus 2011a). Most of the smaller lakes and ponds that existed south of Highway 12 historically have entirely disappeared. The loss of deep, perennial cold water lakes and ponds has reduced the amount of suitable habitat for freshwater and anadromous fish such as Central California Coast coho salmon and Russian River tule perch (Moyle 2002; NMFS 2010; CDFW n.d.), as well as waterbirds such as diving ducks.

One of the biggest changes to the Laguna landscape in terms of percent loss has been the substantial loss of perennial Valley Freshwater Marsh, from approximately 320 ha (~780 acres) historically to just 60 ha today. Much of the area that historically supported freshwater marsh is now comprised of Agriculture and Farmed Wetlands (see Fig. 5-3). In many cases, this conversion was driven by the channelization of the Laguna and its tributaries and the associated drainage of the surrounding wetlands, followed by conversion to agricultural land uses (Miller 1960, Smith 1990; see page 64). The *Healdsburg Enterprise* (1926), for instance, reported that, following construction of a "drainage ditch large enough to carry off all flood and overflow waters" downstream of Ballard Lake, "lands inundated for many years will be cleared of underbrush and tule." Additional loss and degradation of marsh habitat may have been caused by sediment accumulation (estimated to be up to 2 feet between Occidental Road and the Santa Rosa Creek Flood Channel since the 1950s; PWA 2004b) and construction of dikes and berms through the wetland (which restrict drainage and maintain shallowly flooded conditions throughout much of the year).

In addition to this dramatic loss of marsh area, much of the remaining marsh habitat is highly degraded relative to historical conditions. In contrast to the complex and biodiverse perennial marsh habitats that occupied much of the lower Laguna historically, disturbed marshes today are often characterized by a simplified flora dominated by pioneer species, both native (e.g., *Typha* spp.) and non-native (e.g., *Ludwigia hexapetala*; Baye 2008). The loss and degradation of perennial valley freshwater marsh has impacted a wide range of species that use this habitat for foraging, nesting, or cover, including migratory waterfowl, songbirds such as marsh wren and tricolored blackbird, raptors such as northern harrier, and a variety of reptiles and amphibians (Smith 1990, Honton and Sears 2006, Sloop et al. 2007, Sloop and Hug 2009).

Because historical data limitations required that vernal pool habitat be combined with oak savanna, oak woodland, and valley grassland for the change analysis, the changes in extent of vernal pools or the other individual habitat types within this category cannot be quantified. Overall, the extent of the Oak Savanna or Woodland/Vernal Pool Complex/ Valley Grassland land cover category decreased slightly (by ~5%); 30% of this land cover class is today comprised of irrigated pasture or intensively managed hayfields. However, the extent of vernal pool habitat within the study area has likely experienced a much greater decline: visual comparison of 1940s and contemporary aerial photographs (along with modern wetland mapping) reveals numerous areas within the study area where vernal pools were present in the 1940s but are no longer present today. Though many areas were already heavily modified by the 1940s, the historical aerial imagery could be used to develop a minimum estimate of vernal pool loss since the mid-20th century.

On the Santa Rosa Plain as a whole, it is estimated that more than 80% of the historical vernal pool habitat has been eliminated. The major causes of this habitat loss and fragmentation are urban and agricultural development; in addition, degradation of remaining vernal pool habitat has resulted from irrigation, other hydrologic modifications, introduction of non-native species, and other factors (USFWS 2016). Loss of vernal pools on the Santa Rosa Plain has contributed to population declines of a number of species, including four federally and state endangered plants—Sonoma sunshine (*Blennosperma bakeri*), Burke's goldfields (*Lasthenia burkei*), Sebastopol meadowfoam (*Limnanthes vinculans*), and many-flowered navarretia (*Navarretia leucocephala* ssp. *plieantha*)—and the federally endangered and state threatened Sonoma County Distinct Population Segment of the California Tiger Salamander (Smith 1990; USFWS 2005; Honton and Sears 2006; USFWS 2016).

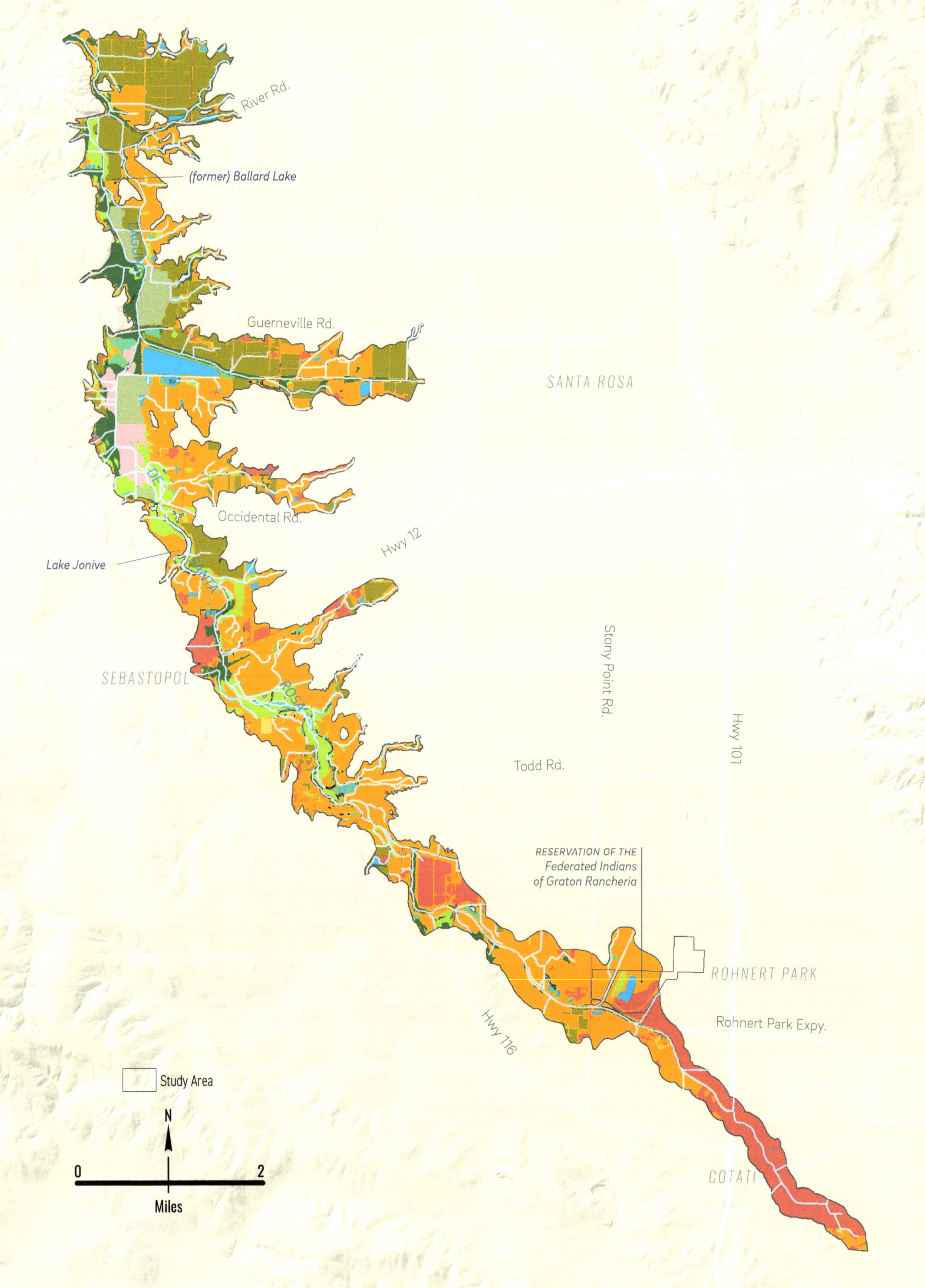
River Rd.
(former) Ballard Lake
LAGUNA
Guerneville Rd.
SANTA ROSA
DE
Occidental Rd.
Hwy 12
Lake Jonive
SANTA
SEBASTOPOL
ROSA
Stony Point Rd.
Hwy 101
Todd Rd.
RESERVATION OF THE
Federated Indians
of Graton Rancheria
ROHNERT PARK
Rohnert Park Expy.
Hwy 116
Study Area
N
0
2
Miles
COTATI

Figure 5-1 (left). Map of modern habitat types and channels within the Laguna de Santa Rosa study area. Land cover data was compiled from NCARI and Sonoma Veg Map data layers.

Figure 5-2 (right). Bar chart shows change in extent of each land cover type within the study area between historical (ca. 1850) and modern (ca. 2015) time periods. Table shows percent change in comparable habitat types between the historical and modern periods.

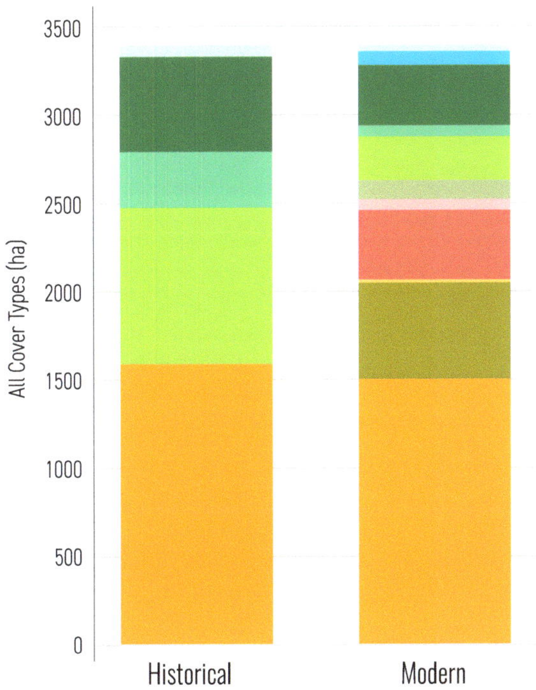

Habitat Types

Habitat Types	ca.1850	ca. 2015	% change
Perennial Freshwater Lake/Pond	53	32	-40%
Wet Meadow	888	234	-74%
Valley Freshwater Marsh	316	61	-81%
Forested Wetland and Mixed Riparian Forest/Scrub	538	339	-37%
Oak Savanna or Woodland/Vernal Pool Complex/Valley Grassland	1585	1625	3%
Other Upland	1	20	1900%
Non-native Aquatic/Emergent Vegetation	0	68	n/a
Storage Pond	0	80	n/a
Farmed Wetland	0	113	n/a
Developed/Disturbed	0	386	n/a
Agriculture	0	423	n/a

Historical and Modern Views of the Laguna. 1942 and 2016 aerial imagery comparing the same location. A) Santa Rosa Creek confluence with the Laguna in 1942. B) Santa Rosa Creek confluence with the Laguna in 2016. C) Mainstem of the Laguna, downstream of the confluence with Bellevue-Wilfred Channel in 1942. D) Mainstem of the Laguna, downstream of the confluence with Bellevue-Wilfred Channel in 2016. Imagery: NAIP.

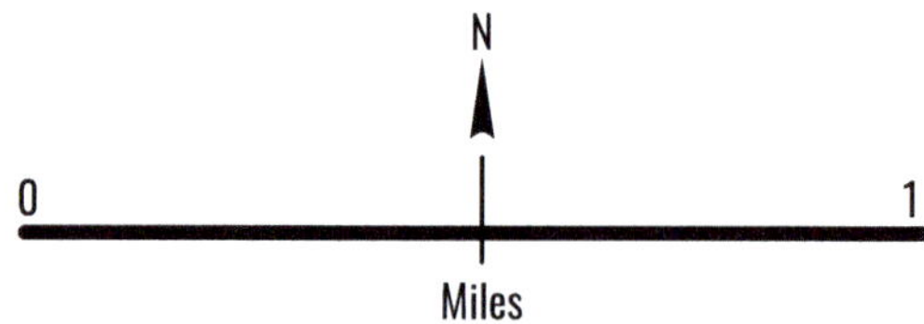

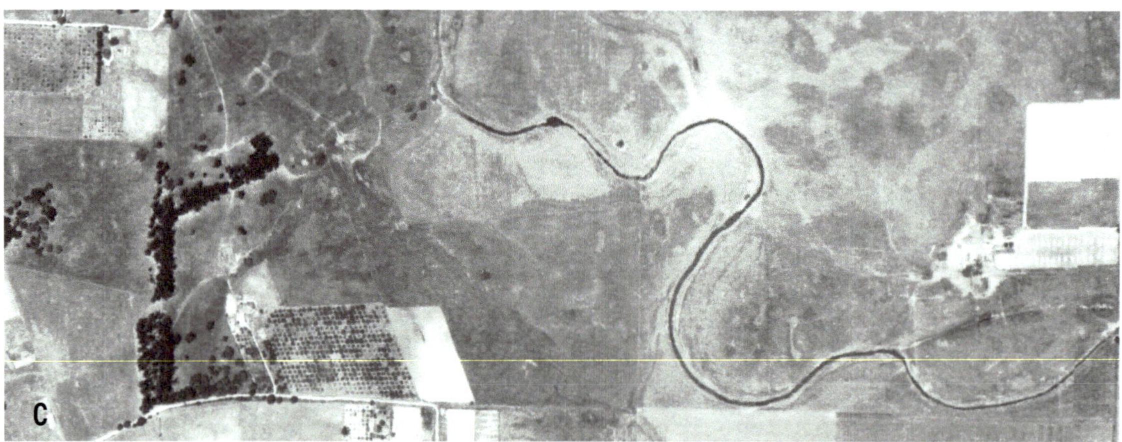

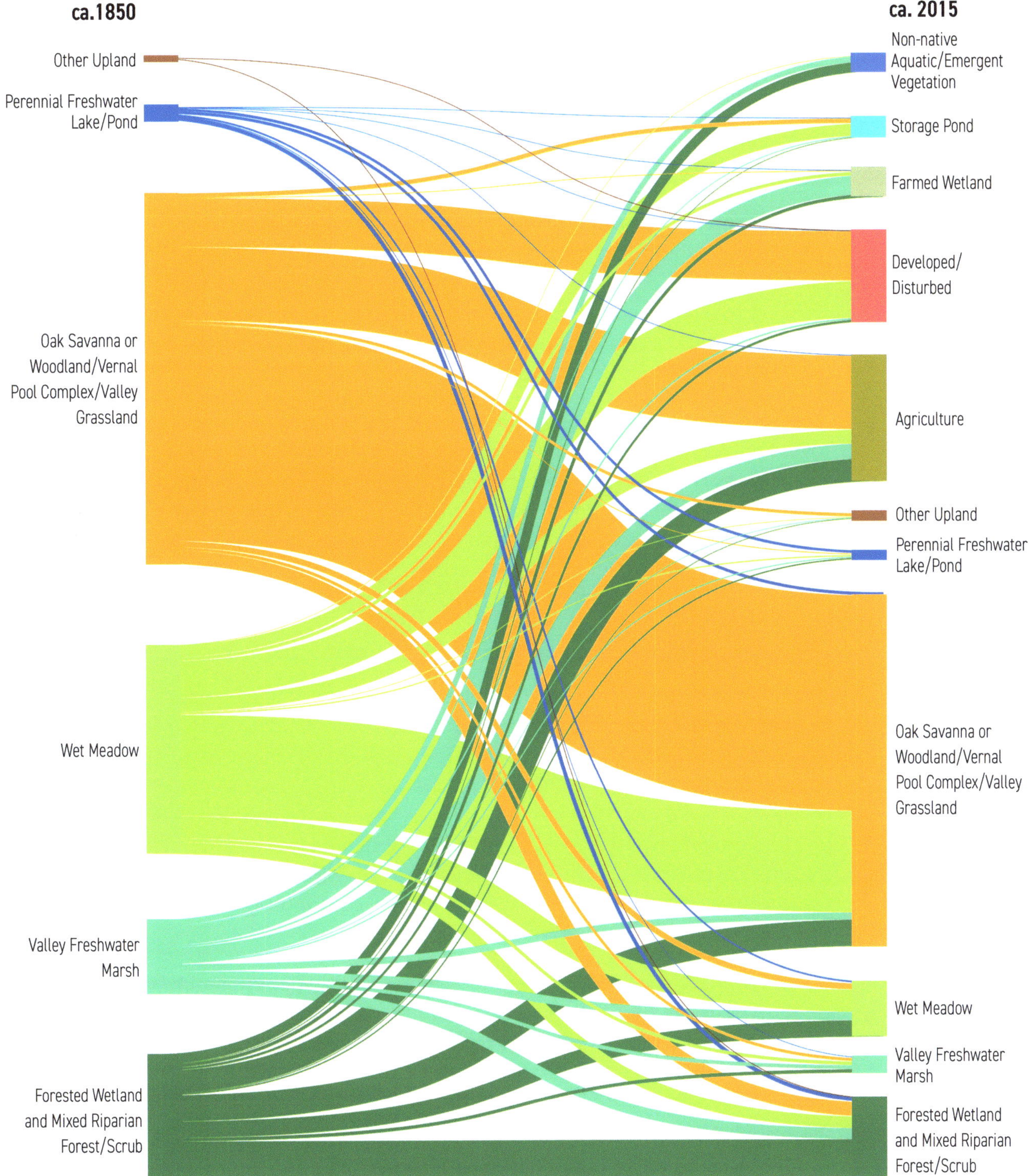

Figure 5-3. Land cover conversion that has occurred within the Laguna over the past two centuries. The bars on the left side represent the proportion of each habitat type present within the study area historically (ca. 1850), while the bars on the right represent the proportion of each habitat type present today (ca. 2015). The lines connecting the left and right sides of the chart illustrate the conversion "pathways" that have occurred for different land cover types since the mid-19th century (e.g., the proportion of Valley Freshwater Marsh that has converted to Agriculture). The thickness of each line corresponds to the total area that has undergone a given type of transformation.

Changes in Habitat Configuration

In addition to overall habitat extent, the configuration of different habitat types has a strong influence on the ability of the landscape to support native species and provide other ecological functions. The size and shape of habitat patches, the number of patches, the degree of connectivity between patches, and the position of habitat patches relative to other land cover types all interact to determine the potential of the landscape to sustain robust native biodiversity in the face of rapid environmental change—in other words, "landscape resilience" (see pages 20-21; Beller et al. 2019).

The configuration of habitats within the Laguna has changed dramatically over the past 150-200 years. These changes include wetland fragmentation, narrowing of riparian corridors, changes in the make-up of habitats adjacent to stream channels, and modifications in the terrestrial zones around wetlands. The following sections quantify changes in these different metrics of habitat configuration, and discuss some of the implications for wildlife support and other ecological functions.

WETLAND FRAGMENTATION

Both wetland and riparian habitats within the Laguna have become fragmented over time as a result of road construction, urban and agricultural development, and other landscape modifications. The loss of large, contiguous areas of habitat has likely impacted wildlife species in a number of ways. Small habitat patches tend to experience more intense edge effects, such as predation and altered abiotic conditions, which may translate into reduced population viability. In addition, the reduced connectivity between patches may hinder dispersal and colonization, contributing to population declines (Fischer and Lindenmayer 2007).

Although there is relatively little information about minimum patch size requirements for particular freshwater wetland species, evidence suggests that the probability of occurrence of certain bird species associated with both valley freshwater marsh and wet meadow is positively correlated with wetland area. Studies of American Bittern in the Midwest and East Coast, for instance, found that they are sensitive to wetland size and only nest in patches of at least 2.5-11 ha (~6.2-27 acres; Brown and Dinsmore 1986; Gibbs and Melvin 1992; Riffell et al. 2001). Virginia Rail, a species of local management concern in the Laguna (Sloop and Hug 2009), has also been shown to have a higher likelihood of occurring in larger wetland patches (Riffell et al. 2001; Richmond et al. 2010).

HISTORICAL and MODERN MARSH PATCHES

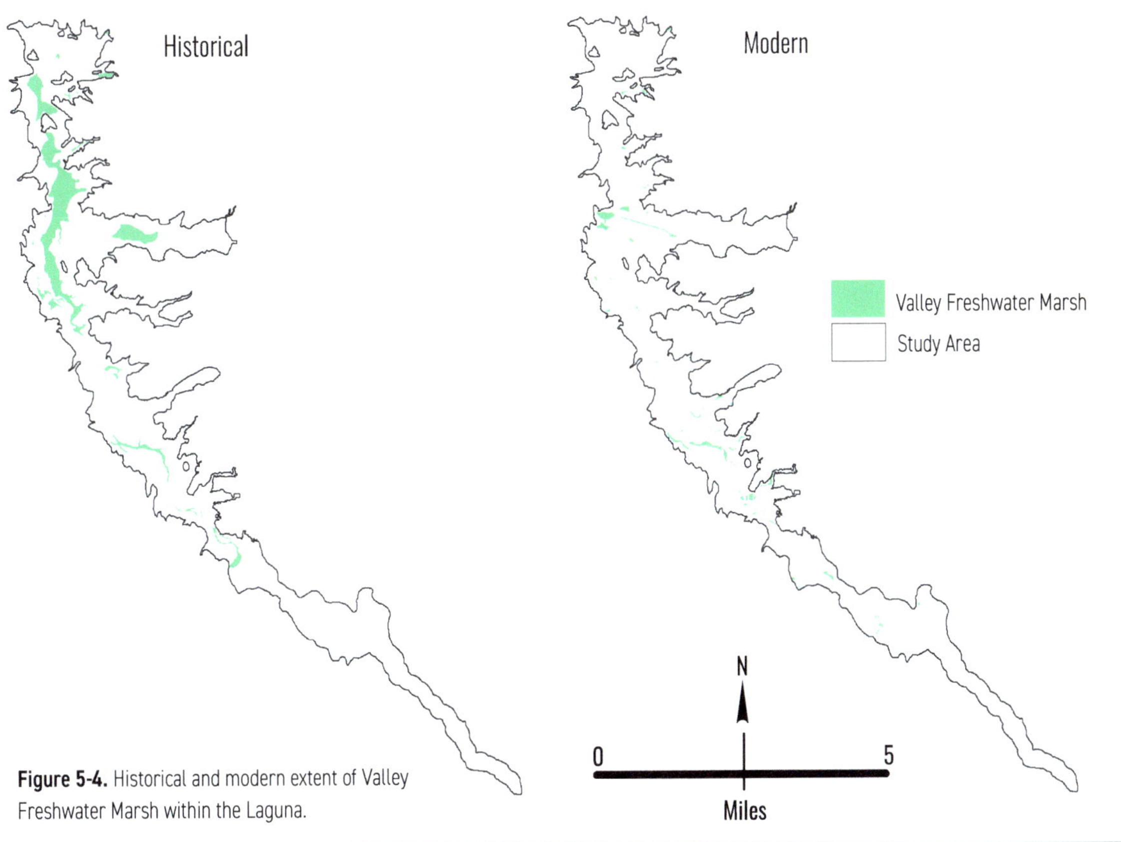

Figure 5-4. Historical and modern extent of Valley Freshwater Marsh within the Laguna.

Figure 5-5. Large marsh patches (>10 ha, 25 ac) historically made up >90% of total marsh area, but make up <20% of the total modern area. Small marsh patches (<1 ha, 2.5 ac) make up >25% of modern marsh area, compared with <1% historically.

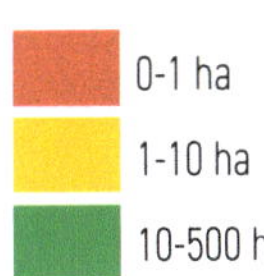

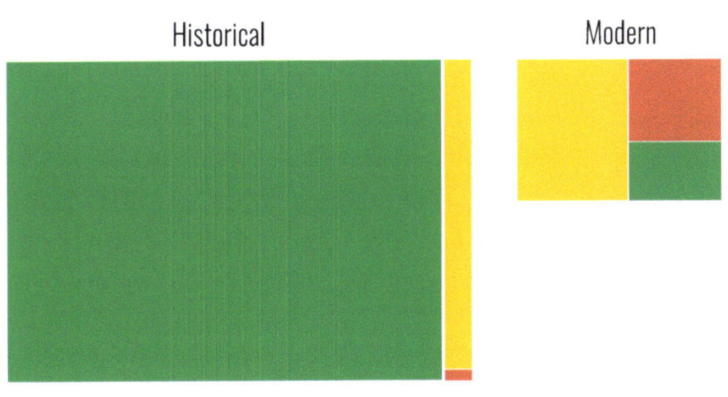

Table 5-2. Marsh area by distance to nearest large (>10 ha, 25 ac) patch. Historically, the vast majority of the marsh area (>95%) was within 500 m (~0.3 miles) of the nearest large patch, while ~70% of modern marsh patches are, on average, greaterr than 500 m apart.

Distance to nearest large patch (m)	Area of marsh habitat (ha)	
	Historical	*Modern*
0-500	303	18
500-1000	1	2
1000-10000	12	34
>10000	0	6
Total	**316**	**61**

Within the Laguna, there has been a substantial loss of the large valley freshwater marsh patches that existed historically (Fig. 5-4). Patches greater than 10 ha (~25 acres) historically made up more than 90% of total marsh area, while today those large patches make up less than 20% of the total; in contrast, over 25% of the contemporary marsh area is comprised of small patches less than 1 ha (~2.5 acres), compared with less than 1% historically (Fig. 5-5). Connectivity between marsh patches, as measured by nearest large neighbor distance, has also decreased substantially: historically, the vast majority of the marsh area (>95%) was within 500 m (~0.3 miles) of the nearest large patch (defined as a patch > 10 ha), while today marsh patches are much further away from each other on average – over 65% of the contemporary marsh area is in patches further than 1,000 m (~0.6 miles) from the nearest large patch (Table 5-2, previous page).

Wet meadow patch size has also decreased relative to its historical distribution, though the change has been less dramatic than for freshwater marsh (Fig. 5-6). Historically, nearly all (97%) of the wet meadow area was comprised of large (>10 ha) patches, while large patches today account for about 77% of the total area (Fig. 5-7). In addition, the largest remaining wet meadow patch (~90 ha, ~220 ac) is much smaller than the largest historical patch (>290 ha, ~720 ac). Connectivity between wet meadow patches has likewise decreased: the percent of patches within 500 m of the nearest large (>10 ha) patch has declined from 99% to 83% (Table 5-3).

HISTORICAL and MODERN WET MEADOW

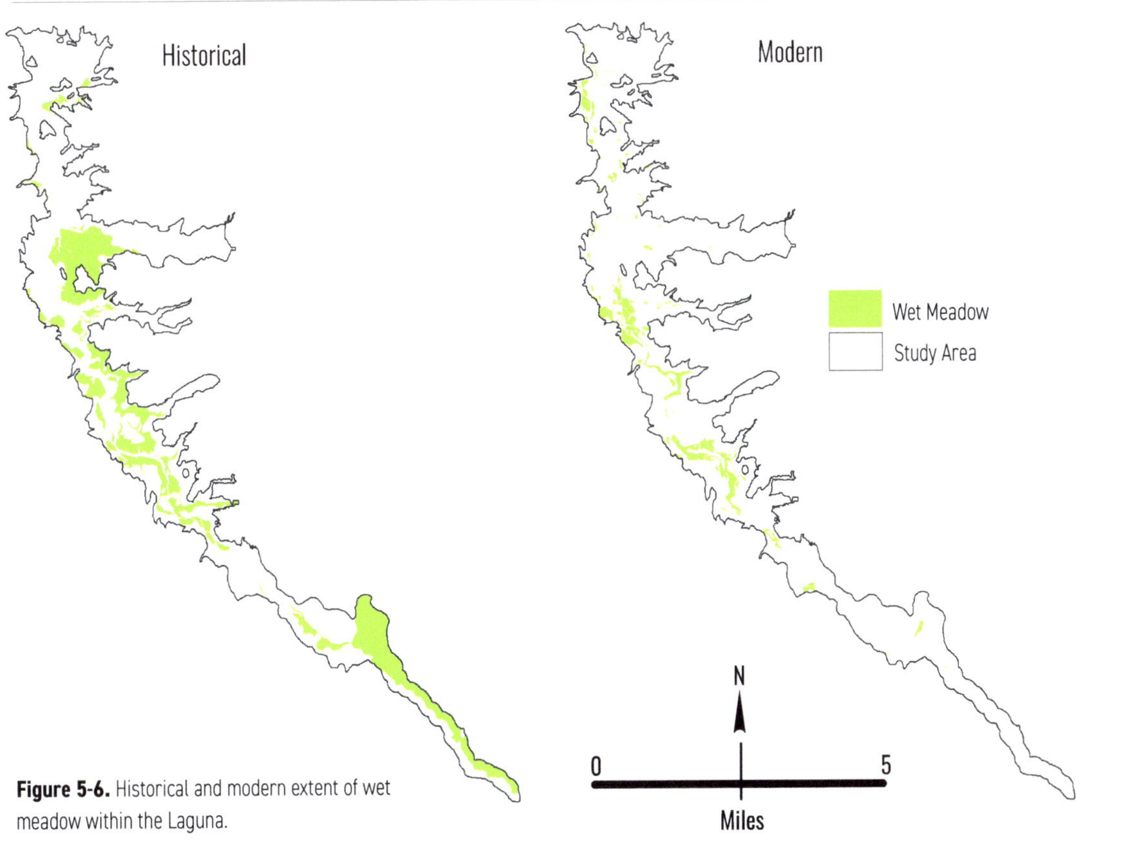

Figure 5-6. Historical and modern extent of wet meadow within the Laguna.

Figure 5-7. Large wet meadow patches (>10 ha, 25 ac) historically made up 97% of total wet meadow area, but make up just 77% of the total area today. Small wet meadow patches (<1 ha, 2.5 ac) make up 6% of modern wet meadow area, compared with <1% historically.

Table 5-3. Wet meadow area by distance to nearest large (>10 ha, 25 ac) patch. Connectivity between wet meadow patches has decreased: the percent of patches within 500 m (~0.3 miles) of the nearest large patch has declined from 99% to 83%.

Distance to nearest large patch (m)	Area of wet meadow habitat (ha)	
	Historical	***Modern***
0-500	878	194
500-1000	6	16
1000-10000	5	23
Total	**888**	**234**

NARROWING OF RIPARIAN CORRIDORS

As with wetlands, the configuration of riparian habitats in the Laguna has also changed considerably over time (Fig. 5-8). However, whereas the most notable change in wetland configuration has been fragmentation of formerly contiguous habitat, the dominant change in the configuration of riparian habitats has been the narrowing of the wide riparian forests that once occupied much of the northern part of the Laguna (Fig. 5-9). Historically, 43% of Forested Wetland and Riparian Forest/Scrub in the Laguna was between 100-600 m in width, and 57% was less than 100 m in width (Fig. 5-10, next page). Today, in contrast, riparian habitats 100-600 m wide make up just 10% of the total area, while habitats less than 100 m wide make up nearly 90% of the total (Table 5-4). Not surprisingly, the narrowing of riparian corridors in the Laguna has been accompanied by a shift towards smaller riparian patches, though the current distribution of riparian habitats is still dominated by relatively large (>10 ha) patches (Fig. 5-10).

The shift towards narrower riparian habitats along the Laguna has likely decreased habitat suitability for a number of wildlife species. For instance, in the Russian River area, mammalian predators such as striped skunk *(Mephitis mephitis)*, raccoon *(Procyon lotor)*, bobcat (*Lynx rufus*), coyote *(Canis latrans)*, and gray fox (*Urocyon cinereoargenteus)* were found to use wide (360-1450 m) riparian areas more frequently than narrow (11-28 m) areas, and the relative proportion of native versus non-native mammalian predator species was higher in wide riparian areas (Hilty and Merenlender 2004). Similarly, a study of songbird presence at Point Reyes National Seashore and the Golden Gate National Recreation Area found that wider riparian habitats were more likely to be occupied by species such as Warbling Vireo *(Vireo gilvus)*, Common Yellowthroat (*Geothlypis trichas*; a species of local management concern in the Laguna; Sloop and Hug 2009), and Swainson's Thrush *(Catharus ustulatus)*; the mean width of riparian habitats occupied by Warbling Vireo was 82 m (Holmes et al. 1999). Yellow-billed Cuckoo (*Coccyzus americanus*), another songbird species that bred in the Laguna historically but has been extirpated (see page 50), requires riparian habitat at least 100 m wide and optimally greater than 600 m wide; habitat 100-200 m wide is considered marginal, and habitat 200-600 m wide is considered suitable (Laymon and Halterman 1989).

Unlike herbaceous wetland habitats in the Laguna, which have become highly fragmented, the connectivity of riparian corridors appears to have remained relatively high (Table 5-4). Based on the nearest neighbor analysis, the proportion of riparian habitat within 500 m of the nearest large patch has decreased somewhat, but this appears to be due primarily to the overall decrease in riparian width and patch size, rather than to fragmentation of existing patches. In fact, riparian forest cover has expanded into some areas

HISTORICAL and MODERN RIPARIAN CORRIDORS

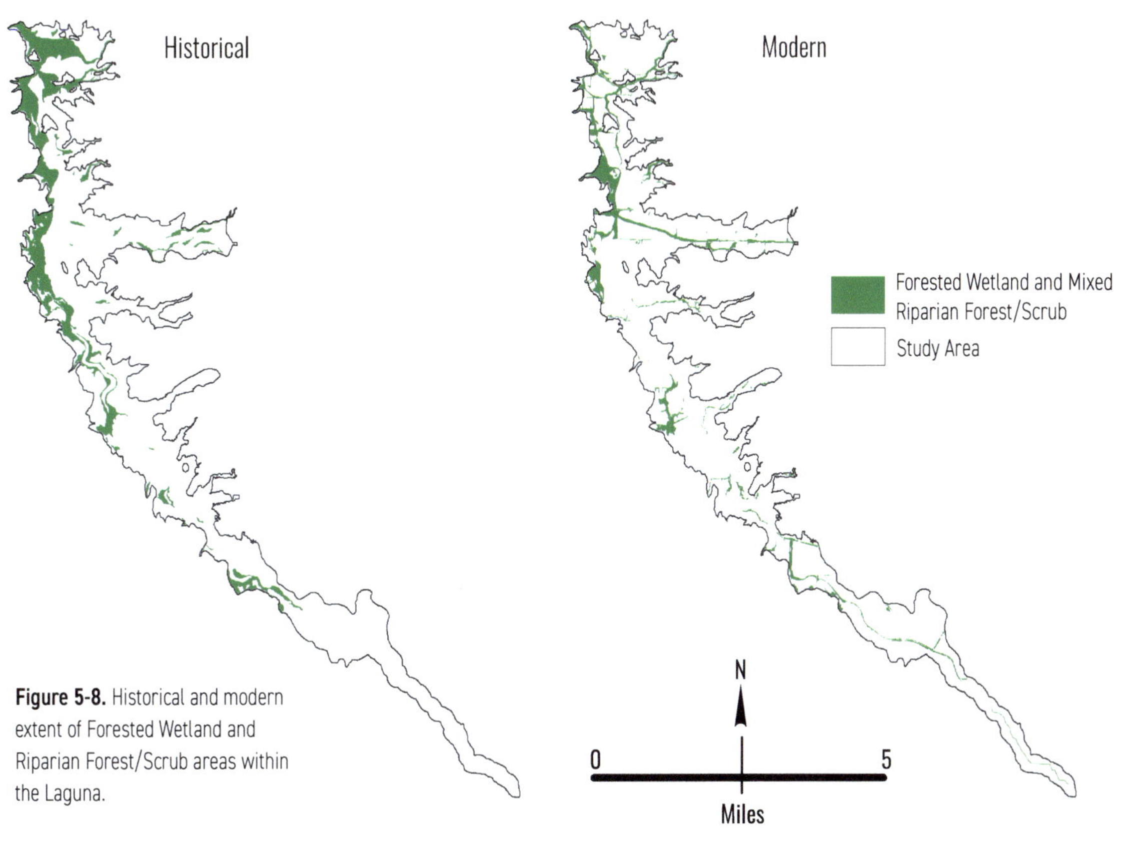

Figure 5-8. Historical and modern extent of Forested Wetland and Riparian Forest/Scrub areas within the Laguna.

Figure 5-9. Historical and modern riparian patch size.

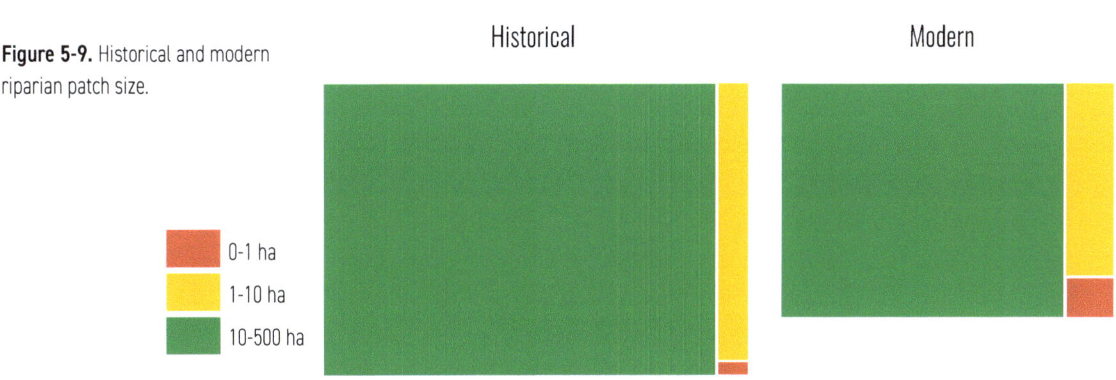

Table 5-4. Riparian area by distance to nearest large (>10 ha, 25 ac) patch. The proportion of riparian habitat within 500 m (~0.3 miles) of the nearest large patch has decreased somewhat, likely due to the overall decrease in riparian width and patch size.

Distance to nearest large patch (m)	Area of riparian habitat (ha)	
	Historical	*Modern*
0-500	513	308
500-1000	11	17
1000-3000	13	14
Total	**538**	**339**

where it was not documented historically, most notably along lower Santa Rosa Creek and in the southern portion of the study area around Rohnert Park and Cotati.

In addition to decreasing habitat suitability, the narrowing of riparian corridors along the Laguna has also likely impacted a number of ecosystem services historically provided by riparian forests, including runoff filtration, nutrient removal, sediment storage, and temperature regulation. For instance, a literature review examining how riparian buffer width affects water quality and other ecosystem services found that buffers greater than ~30 m wide were needed to maximize services such as nitrate removal, sediment trapping, temperature regulation, and large woody debris inputs (Sweeney and Newbold 2014).

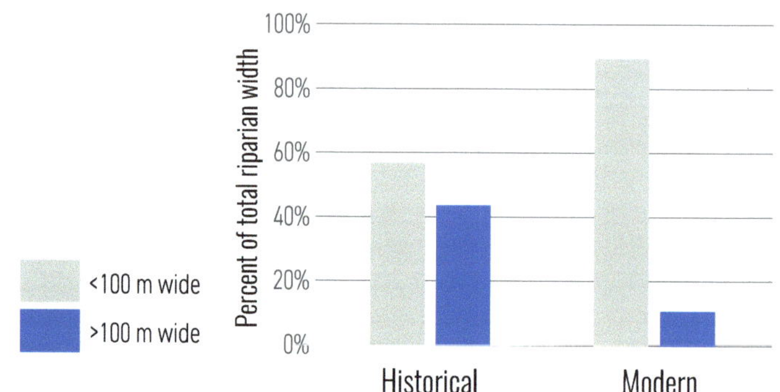

Figure 5-10. Proportion of narrow and wide riparian widths. Historically, 43% of Forested Wetland and Riparian Forest/Scrub in the Laguna was greater than 100 m in width; today, in contrast, riparian habitats greater than 100 m wide make up just over 10% of the total.

Table 5-5. Length of riparian habitat by width class.

Riparian habitat width (m)	Length of riparian habitat (km)	
	Historical	***Modern***
0-100	21	51
100-600	16	6
Greater than 600	<1	<1
Total	**37**	**57**

Field and riparian vegetation along lower Irwin Creek. Photo: SFEI.

CHANNEL-HABITAT ADJACENCY

The composition of floodplain habitats immediately adjacent to stream channels, referred to here as "channel-habitat adjacency," is another informative measure of changes in habitat configuration within the Laguna. Many wildlife species, including various aquatic birds and reptiles, use multiple habitat types along a wet-dry gradient under the right conditions, and thus the presence of intact floodplain habitats (both wetland and terrestrial) adjacent to stream channels can support those different life history options. For instance, in a study of western pond turtle (*Actinemys marmorata*) habitat use in Northern California, Reese (1996) found that turtles frequently used wetland and terrestrial habitats adjacent to streams and lakes. Lentic off-channel habitats, such as marshes, ponds, sloughs, and vernal pools, likely provide a number of functions for pond turtles, including refuge from high velocity flows during the winter and refuge from aquatic predators; use of these habitats was found to be especially frequent among juvenile turtles (Reese 1996).

In addition to its function as wildlife habitat, the presence of wetland and riparian areas adjacent to stream channels provides a number of other ecosystem services. Primary production within freshwater marshes and seasonally inundated habitats provides food inputs to adjacent channels, which helps support salmonids and other native fishes (Henning et al. 2006, Opperman et al. 2017). Wetlands and riparian corridors filter stormwater runoff, buffering stream channels from external inputs of nutrients, sediment, and pollutants and helping to maintain water quality downstream (Mitsch and Gosselink 1993, Teels et al. 2006, Kaushal et al. 2008). In some circumstances, fringing wetlands can also reduce the levels of nutrients (nitrogen) in streams, lakes, and ponds by reducing the presence of organisms in the water column that fix nitrogen from the atmosphere (Tomasko et al. 2016). Riparian forests shade adjacent aquatic habitats, which helps to regulate water temperature, and provide inputs of leaf litter, large woody debris, and macroinvertebrates (Wenger and Fowler 2000, Teels et al. 2006).

Among stream channels within the study area, the composition of floodplain habitats has changed considerably over the past two centuries (Fig. 5-11). One of the most notable changes has been the decrease in the proportion of herbaceous wetland types (valley freshwater marsh and wet meadow) adjacent to stream channels: historically, these two wetland types made up approximately 39% of the floodplain habitats adjacent to channels (as measured by channel length), while today they comprise just 18% of adjacent habitats. The magnitude of this change is even more pronounced when looking just at the mainstem channel alone: the fraction of the mainstem channel bordered by valley freshwater marsh and wet meadow has decreased from approximately 48% to 15%, while the fraction of the channel bordered by forested wetland and riparian forest/scrub has increased from 40% to 65% (Fig. 5-12).

HABITATS ADJACENT TO ALL CHANNELS

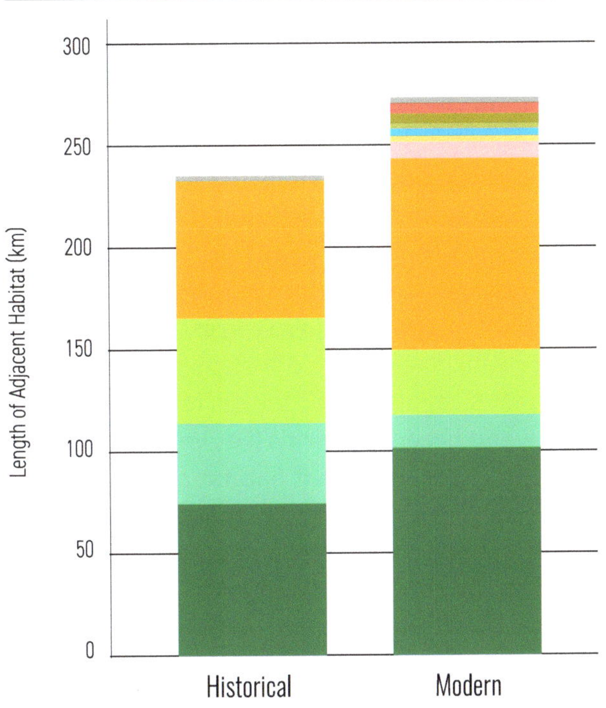

Figure 5-11. Length of all channels by adjacent habitat type.

HABITATS ADJACENT TO MAINSTEM LAGUNA

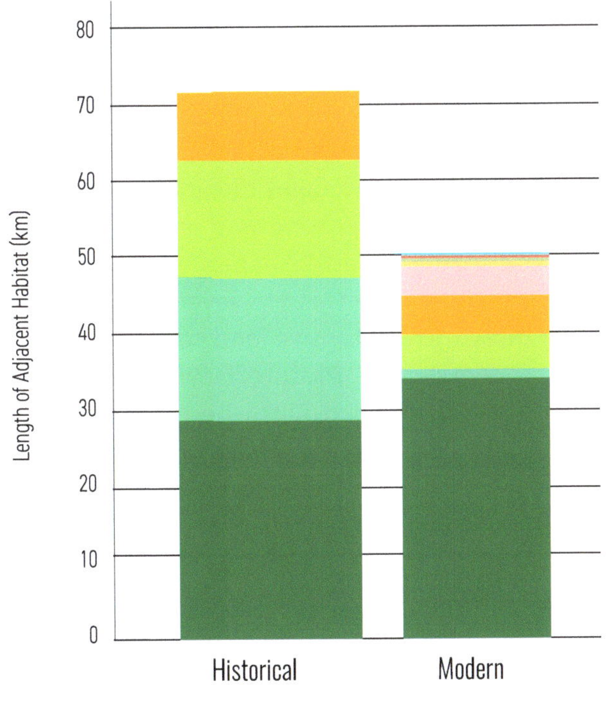

Figure 5-12. Length of mainstem by adjacent habitat type.

TERRESTRIAL ZONES AROUND WETLANDS

Just as the presence of wetlands adjacent to stream channels is important for wildlife support and other functions, the presence of a wide terrestrial buffer zone around wetland and aquatic habitats is important for many species. This is particularly true for semiaquatic species, including many reptiles and amphibians, which require both terrestrial and wetland habitats for different portions of their life history (e.g., wetlands for breeding or foraging and upland habitat for overwintering). For example, Bulger et al. (2003) found that California red-legged frogs (*Rana draytonii*) require terrestrial buffer zones at least 100 m wide around aquatic habitats, and that the terrestrial zones should include dense herbaceous vegetation or shrubs for protective cover. Likewise, upland habitat between breeding ponds (i.e., vernal pools or other suitable wetland habitats) provides important habitat connectivity for California Tiger Salamanders (USFWS 2016). A meta-analysis looking at 65 species of reptiles and amphibians found that the core habitat for these species includes terrestrial zones ranging from approximately 140-290 m (~150-320 yds) wide adjacent to aquatic and wetland habitats (Semlitsch and Bodie 2003).

Figure 5-13 shows the extent of a terrestrial buffer zone approximately 140 m wide (the minimum width identified by Semlitsch and Bodie 2003) around contemporary aquatic and wetland habitats within the study area. Though the historical composition of this terrestrial buffer zone was not analyzed, it is assumed that historically this zone would have provided suitable terrestrial habitats for a wide range of species. Figure 5-14 shows the current area of terrestrial habitats and other land cover types within this zone. Valley Grassland occupies the greatest area within the terrestrial buffer zone, and presumably provides suitable terrestrial habitat for a range of semiaquatic species. However, the next three most extensive land cover types–Agriculture, Pasture/Hayfield, and Developed/Disturbed–likely represent marginal or unsuitable habitat for many species.

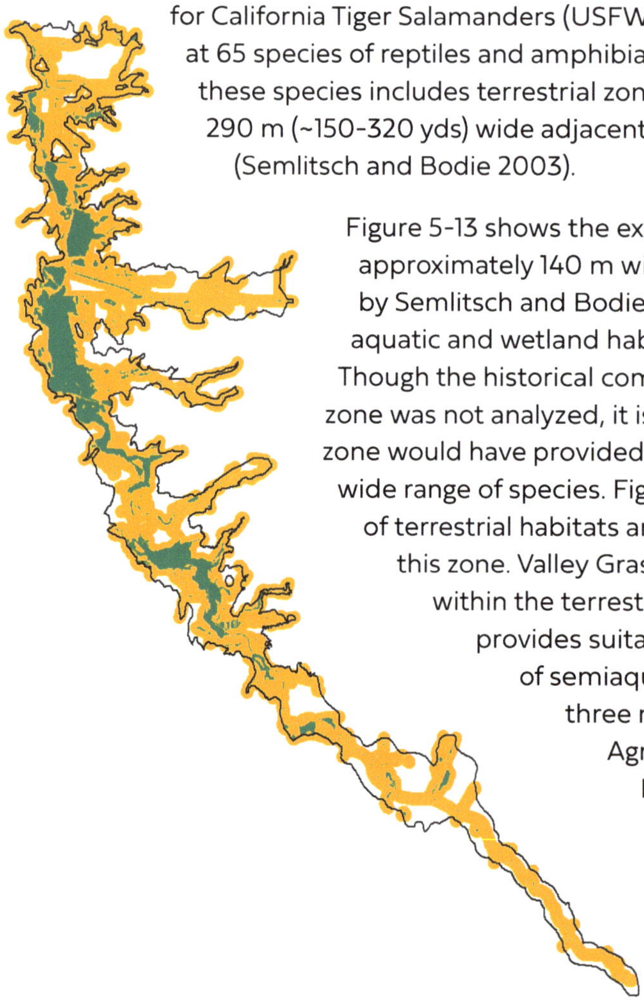

Figure 5-13. Terrestrial buffer (gold) around aquatic and wetland habitats (teal) within the Laguna Study Area.

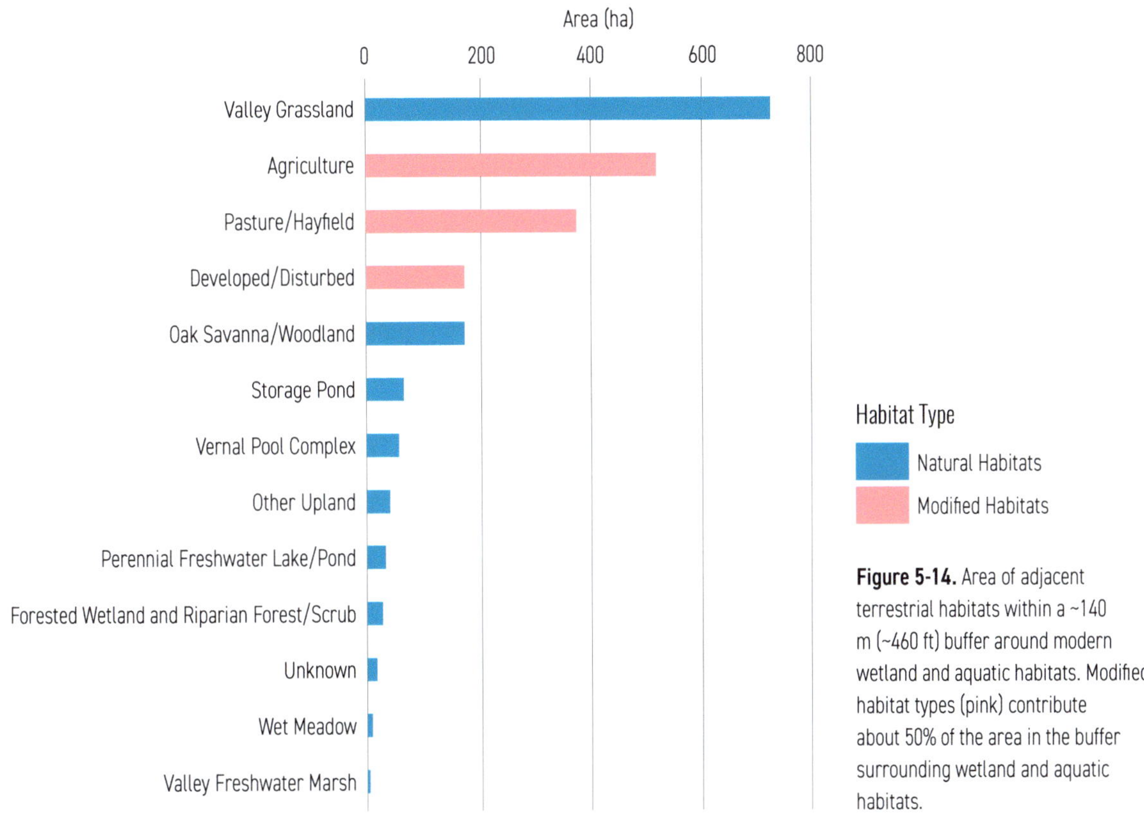

Figure 5-14. Area of adjacent terrestrial habitats within a ~140 m (~460 ft) buffer around modern wetland and aquatic habitats. Modified habitat types (pink) contribute about 50% of the area in the buffer surrounding wetland and aquatic habitats.

Pasture surrounding seasonal wetlands in the Laguna de Santa Rosa. Photo: SFEI.

CHANGES IN CHANNEL NETWORK

In addition to changes in land cover extent and habitat configuration, there have also been significant changes in the configuration of the channel network within the study area over the past 200 years (Fig. 5-15). In general, channel modifications were intended to increase drainage efficiency, reduce flooding extent, and increase available land for agricultural or urban use.

Along the Laguna mainstem, straightened channels were constructed through a number of areas that were historically characterized by broad wetland complexes and a network of shallow, poorly defined sloughs. For example, in 1927-8 a series of ditches were constructed downstream of Ballard Lake, which substantially lowered water levels in the lake and other areas of the Laguna further upstream (*Healdsburg Enterprise* 1926; *Healdsburg Tribune* 1927; *The Press Democrat* 1928). Further upstream, a segment of the Laguna extending approximately 13 km (8 mi) from Occidental Road to a point about 1.6 km (1 mi) downstream of Guerneville Road was channelized in 1966 (Miller 1960; Beach 2002; PWA 2004a; Cummings 2004). In the southern portion of the study area around Cotati, the meandering channel that existed historically was replaced by a straightened flood channel. Overall, the length of the Laguna mainstem channel has decreased by about 19% as a result of channel simplification (i.e., straightening and the conversion of some reaches from multi-threaded to single threaded).

Major changes have also occurred along tributary creeks. As urban and agricultural development extended across the Santa Rosa Plain during the late 19th and 20th centuries, portions of many tributary channels were leveed and straightened to control flooding and promote drainage efficiency. For example, a flood control channel was constructed along lower Santa Rosa Creek in the 1960s, converting the downstream portion from a complex multi-threaded morphology to a straightened, single-threaded channel. Tributaries like Copeland and Crane creeks in the southern part of the Laguna, which historically terminated in a series of distributary channels on the Cotati Plain upstream of the Laguna mainstem, were lengthened to connect with the Laguna and increase drainage efficiency (Dawson and Sloop 2010). Modifications that eventually resulted in the construction of the Bellevue-Wilfred Flood Channel began in the 1920s with the excavation of a series of ditches by the local drainage district (*The Press Democrat* 1920).

These and other changes have resulted in an approximately 30% increase in the length of tributary channels within the study area. Channelization of tributaries has altered hydrologic and sediment delivery patterns within the study area (see pages 23 and 30) and decreased the availability and complexity of both in-channel and floodplain habitats. These changes have negatively impacted species such as the federally endangered California

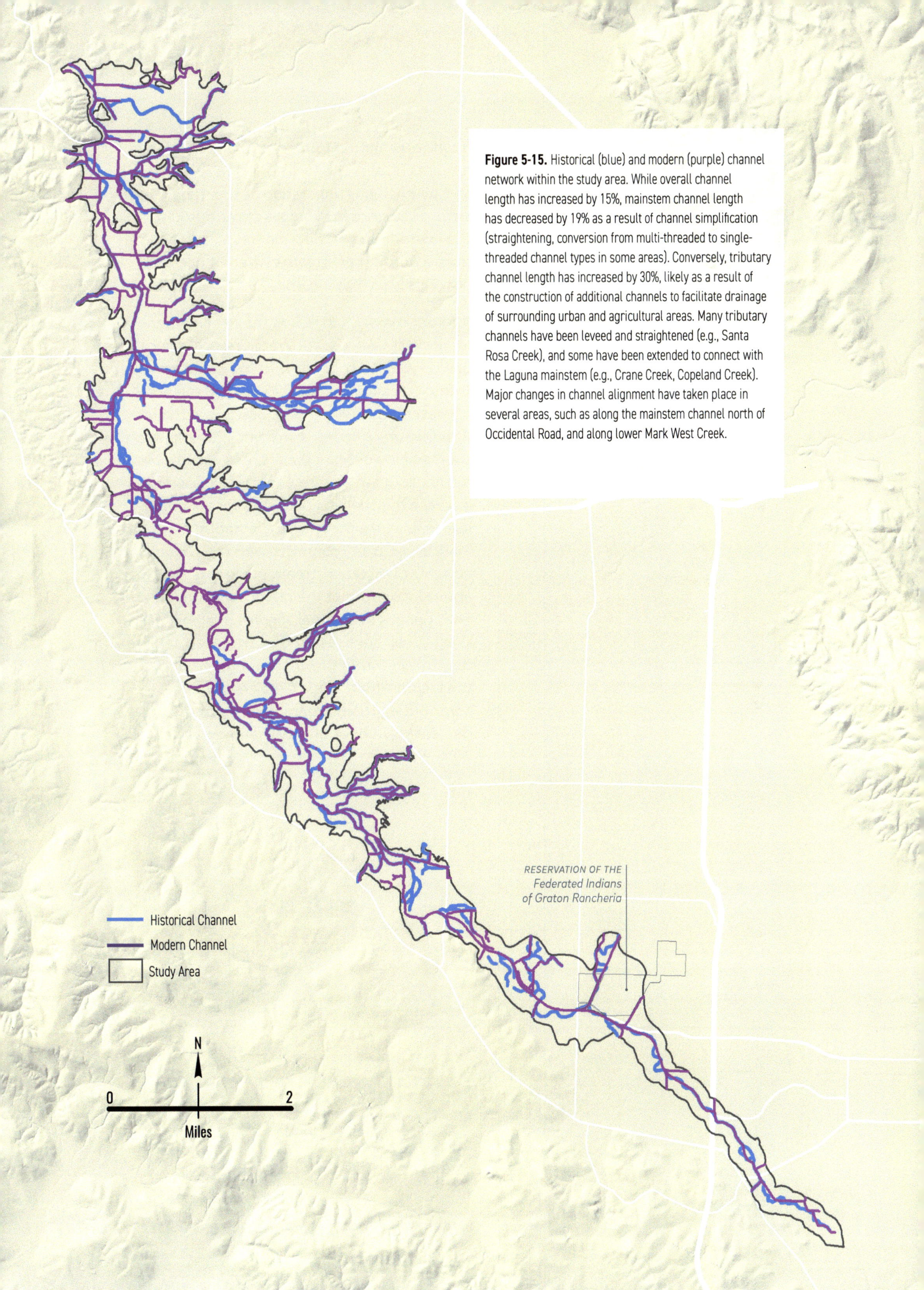

Figure 5-15. Historical (blue) and modern (purple) channel network within the study area. While overall channel length has increased by 15%, mainstem channel length has decreased by 19% as a result of channel simplification (straightening, conversion from multi-threaded to single-threaded channel types in some areas). Conversely, tributary channel length has increased by 30%, likely as a result of the construction of additional channels to facilitate drainage of surrounding urban and agricultural areas. Many tributary channels have been leveed and straightened (e.g., Santa Rosa Creek), and some have been extended to connect with the Laguna mainstem (e.g., Crane Creek, Copeland Creek). Major changes in channel alignment have taken place in several areas, such as along the mainstem channel north of Occidental Road, and along lower Mark West Creek.

California Freshwater Shrimp (female with eggs during shrimp survey on creek in Marin County). Photo: Alex Iwaki, NPS.

freshwater shrimp (*Syncaris pacifica*), which requires undercut banks, exposed root material, and overhanging vegetation at the margins of streams; channelization has contributed to the extirpation of California freshwater shrimp from a number of streams where it occurred historically (USFWS 2011).

Another major change was the re-alignment of Mark West Creek. Mark West Creek historically flowed west to connect with the Laguna near Trenton, approximately 0.75 km (0.5 mi) north of present-day River Road. Starting in the late 19th century, a series of modifications shifted the course of lower Mark West Creek progressively further south in an effort to control flooding and increase the availability of land for agricultural use. In its current alignment, established in 1963, Mark West Creek connects with the Laguna about 0.75 mi downstream of Guerneville Road, about 2 mi south of the historical confluence. The changes in channel alignment along Mark West Creek have resulted in increased sediment deposition within the channel and around the Laguna-Mark West Creek confluence, contributing to upstream flooding and decreasing habitat quality (Baumgarten et al. 2014, 2017).

Introduction of Invasive Species

Most areas of the Laguna have been colonized by non-native plant species, many of which do not cause ecosystem-level changes (Honton and Sears 2006). However, several invasive plant species of concern are prevalent within the Laguna (Table 5-6). Notable species of management concern include Himalayan blackberry *(Rubus armeniacus)*, perennial pepperweed *(Lepidium latifolium)*, and invasive *Ludwigia* species.

These problematic non-native species can form self-sustaining populations and overtake entire areas of the Laguna, displacing even formerly common native species (IUCN 2000). For example, in riparian areas where Himalayan blackberry establishes, it edges out a formerly mixed understory of native shrubs and grasses, diminishing plant and animal

Table 5-6. Invasive plant species of high concern in the Laguna that are likely to compromise restoration efforts if no weed management is planned. These species are currently of the highest management concern (adapted from Honton and Sears 2006, W Trowbridge, B Grewell pers. comm). However, new and unexpected invasions are possible. Monitoring and management of these and future invaders is advisable.

Botanical Name	Common Name	Typical Habitat Type
Aegilops triuncalis	Barbed goatgrass	Grasslands and pastures
Agrostis stolonifera	Creeping bent grass	Wetlands, riparian areas
Alisma lanceolatum	Lanceleaf water plantain	Wetlands
Arundo donax	Giant reed, arundo	Riparian areas
Centaurea solstitialis	Yellow starthistle	Grasslands and disturbed areas
Cortaderia jubata	Pampas grass, jubata grass	Disturbed areas, including ponds and stream banks
Eucalyptus spp.	Eucalyptus species	Riparian, wetlands, grasslands
Festuca perennis	Italian ryegrass	Grasslands, wetlands, vernal pools
Genista monspessulana	French broom	Oak woodlands, grasslands
Holcus lanatus	Velvet grass	Riparian areas and grasslands
Iris pseudacorus	Yellow flag Iris	Riparian, wetland
Lepidium latifolium	Perennial pepperweed	Marshes, riparian areas, wetlands, grasslands
Ludwigia hexapetala	Uruguayan primrose-willow, ludwigia	Aquatic habitats, wetlands
Lythrum salicaria	Purple loosestrife	Wetlands and riparian areas
Mentha pulegium	Pennyroyal	Wetland edges
Myriophyllum aquaticum	Parrotfeather	Wetlands and riparian areas
Phalaris aquatica	Harding grass	Wetlands, riparian areas, and grasslands
Phalaris arundinacea	Reed canary grass	Wetlands and riparian areas
Rubus armeniacus	Himalayan blackberry	Wetlands and riparian areas, disturbed areas
Sesbania punicea	Scarlet wisteria tree	Riparian areas
Taeniatherum caput-medusae	Medusa-head	Grasslands

biodiversity. Perennial pepperweed causes problems in newly disturbed areas, such as restoration sites, and can form dense stands, precluding the establishment and spread of native herbaceous species.

Non-native plants and animals have been introduced to the Laguna over time, both intentionally and unintentionally. Human activities increase non-native plants and animals on the landscape in a number of ways, including through novel introductions and re-introductions (Turbelin et al. 2017); ground and water disturbances that open up areas for colonization such as road cuts, new construction sites, restoration sites, and dredging sites that provide opportunities for non-native species to colonize; and increases in loads and concentrations of nutrients from agricultural and urban areas that change conditions in ways that can favor non-native species (CBD 2014; Essl et al. 2015). In a vicious cycle, some species, once established, contribute to changes in ecosystem processes that create and sustain conditions conducive to their own survival, such as increasing sedimentation, slowing water velocities, changing chemical properties of substrates, and altering food web dynamics (Parker et al. 1999, Solé et al. 2002, Ehrenfeld 2010). Disturbances caused by invasive species rank among the most pervasive of human-caused ecosystem change (Vitousek et al. 1997, Pimentel et al. 2005, Butchart et al. 2010, Vilà et al. 2011), and their management will be a primary component of successful restoration in the Laguna (see pages 138-141 for invasive species management recommendations).

Monitoring for invasive *Ludwigia* spp. in the Laguna de Santa Rosa. Photo: Brenda Grewell.

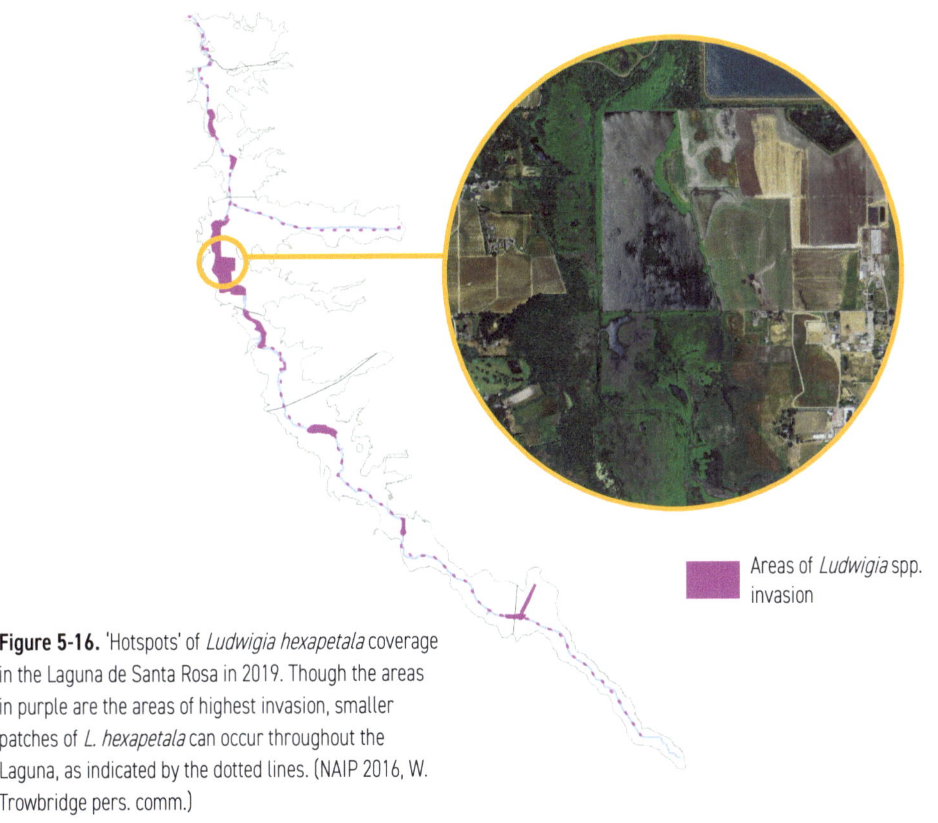

Figure 5-16. 'Hotspots' of *Ludwigia hexapetala* coverage in the Laguna de Santa Rosa in 2019. Though the areas in purple are the areas of highest invasion, smaller patches of *L. hexapetala* can occur throughout the Laguna, as indicated by the dotted lines. (NAIP 2016, W. Trowbridge pers. comm.)

FOCUS ON *LUDWIGIA HEXAPETALA*

The dominant invasive *Ludwigia* species in the Laguna, *Ludwigia hexapetala* (Grewell et al. 2019), covers an estimated 150-300ac of the Laguna, about 2-4% of the study area (Fig. 5-16). This invasion is symptomatic of ecosystem changes in the Laguna watershed that favor its growth and spread, including increased sedimentation, nutrients, and summer water inputs (Meisler 2008). The widespread presence of invasive *Ludwigia* species in the Laguna has been of high management concern for the past few decades (Ludwigia Task Force 2004; Honton and Sears 2006, Sloop et al. 2007, Meisler 2008, Grewell and Futrell 2009, Grewell et al. 2016b). Within invaded areas, invasive *Ludwigia* spp. create several ecological problems, including degradation of habitat by displacement of native plants, reduction of open water and wetland habitats for waterfowl and fish; degradation of water quality by reducing dissolved oxygen (DO) as they decompose seasonally, alteration of and contribution to nutrient cycling (N, P, C); increasing flood risk by displacement of channel water capacity, slowing water velocities; and contribution to hyperaccumulation of sediments (Grewell et al. 2016a, 2019). It also can pose a public health risk by impeding the effectiveness of mosquito vector control activities (Sloop et al. 2007, Meisler 2008). Other risk cofactors for excess invasive *Ludwigia* spp. growth include shallow water depth, lack of riparian cover, low flows, altered flow regime, and high water temperature (Dodds et al. 2002, Sloop et al. 2007).

Water primroses and water primrose-willows (*Ludwigia* spp.) are among the worst invasive wetland plant species in the world (Thouvenot et al. 2013). Two Ludwigia species are aggressive weeds in the Laguna and the greater Russian River Watershed: *L. hexapetala* (Uruguayan primrose-willow) and *L. peploides* subsp. *montevidensis* (creeping water primrose). These emergent plants are rooted in sediment, and produce roots along buoyant stem nodes.

The most commonly found species in the Laguna, *L. hexapetala*, is well-adapted to fluctuating water levels, and can tolerate terrestrial conditions in transitional zones along waterways (Grewell et al. 2016a). A single plant can be a mat-forming perennial herb in shallow water, or it can transform to a woody sub-shrub (hence "primrose-willow") and become more ascending and erect near and above the water's edge (Grewell et al. 2016b). Due to high growth rates and biomass production, particularly in eutrophic areas, the mats they form can often completely cover formerly open-water areas. These mats exclude light from the subsurface water column, and limit the growth of desirable submersed aquatic plants important to aquatic food webs.

Ludwigia hexapetala. Photo: Brenda Grewell.

Both alien *Ludwigia* taxa in the Laguna have three reproductive modes: asexual fragmentation of shoots, asexual fragmentation of rhizomes; and viable seed production (Grewell et al. 2019). All of these propagules are buoyant and disperse rapidly with water currents to expand the invasion. Conditions common to Laguna waterways, such as variation in water velocities and a seasonal drawdown of water level, have been observed to be conducive to *Ludwigia* growth and spread in the Russian River (Skaer Thomason et al. 2018a, 2018b; Grewell et al. 2019). Control measures for invasive *Ludwigia* and other problematic plants are discussed in Chapter 7.

Summary

The Laguna de Santa Rosa landscape has been heavily altered by land and water use changes over the past two centuries, resulting in the loss, fragmentation, and degradation of native habitats and impairment of critical ecological functions. Overall, historical wetland, riparian, and aquatic habitat area has decreased by 63% within the study area, with large areas of former wetland and riparian areas converted to urban and agricultural land uses. Most of the large, deep lakes that existed historically have been lost or greatly reduced in size and depth. Existing wetland patches are small and highly fragmented relative to historical conditions, while existing riparian corridors are in general much narrower than they were in the past. Many of the perennial and seasonal wetlands that historically existed adjacent to stream channels have been lost, and much of the terrestrial buffer zone around wetland and aquatic habitats is today occupied by urban and agricultural land uses. The channel network has also been altered substantially: large sections of the Laguna mainstem channel were ditched and straightened, and portions of many tributaries were channelized (and in the case of lower Mark West Creek, re-aligned entirely). A number of problematic invasive species, most notably *Ludwigia hexapetala*, have been introduced to the Laguna, displacing native plants and degrading water quality. All of these changes have impacted the ability of Laguna to provide desired ecosystem services and ecological functions, and underscore the need for a landscape-scale, multi-benefit restoration vision. §

Smoke from California fires October, 2017. Photo: NASA.

6 Future Conditions and Climate Change Predictions

In the coming decades, the Laguna will face a variety of risks associated with climate change. Greenhouse gas emissions are projected to cause rising temperatures, shifting wet and dry seasons, increasingly volatile precipitation patterns, changing vegetation, drought, and increasing wildfire risk. These outcomes—which will likely be compounded by resultant changes in flooding patterns, sediment transport, and erosion dynamics—threaten the livelihoods of people and wildlife that depend on the Laguna. Below is a high-level overview of the projected shifts in precipitation dynamics and air temperature due to climate change, and the associated impacts to the Laguna ecosystem.

Rising Temperatures and Extreme Heat Events

Future climate models generally agree that the Bay Area overall, and the Laguna de Santa Rosa specifically, will face rising temperatures due to climate change (Micheli et al. 2009, Wiess et al. 2013). Increases in mean temperature correspond with increasingly frequent extreme heat events (Wiess et al. 2013, Dahl et al. 2019), as temperatures are expected to rise disproportionately in the summer months, with three-day heat waves in Sonoma County increasing in temperature by ~2°C (~4°F) (Pierce et al. 2013b). These events pose a significant risk for human health, especially for the elderly and other at-risk populations (Luber and McGeehin 2008). Increased temperatures will also likely lead to warmer water temperatures, contributing to lower dissolved oxygen levels and higher pollutant toxicity in freshwater systems such as the Laguna — two major threats to aquatic wildlife (Ficke et al. 2007). Uncertainty surrounds how increased temperatures will affect flows in local waterways. Higher temperatures increase evapotranspiration, which can lead to reduced runoff, but precipitation changes may overshadow this effect (Pierce et al. 2013a, Woodhouse and Pederson 2018).

Precipitation Patterns

While changes in overall annual precipitation for the Russian River watershed are uncertain, the seasonality and intensity of rainfall are projected to shift into the future. Wet seasons are likely to become shorter and more intense while dry seasons become longer (Mount et al. 2017). Climatic trends over the last century indicate that California is increasingly fluctuating between drought and extreme wet years (He and Gautam 2016). A range of climate scenarios predict that these fluctuations will become more severe into the future and large flood events will likely become more frequent (Dettinger 2013, Micheli et al. 2016, Mount et al. 2017).

Changes in precipitation will present a complex suite of challenges for cities and landowners neighboring the Laguna. More extreme rainfall events could result in more stormwater, sediment, nutrients, and trash being transported from the surrounding landscape into the Laguna and the Russian River. Concentration of annual rainfall within fewer, more intense events could disproportionately increase overland runoff, and decrease the amount of rainwater that infiltrates and recharges groundwater supplies. Meanwhile, with longer dry seasons and higher summer temperatures, farmers will likely require >10% more water for irrigation and rely increasingly upon this diminished resource (Micheli et al. 2016, Mount et al. 2017). This could result in less tributary flow into the Laguna during the dry season, which adversely affects native vegetation and wildlife.

Shifting Vegetation Patterns

Changes in temperature and precipitation patterns will likely result in shifting vegetation in the region. With warmer temperatures and more pronounced summer droughts, drought-tolerant vegetation such as chamise chaparral and coast live oak will likely expand in the Russian River Basin (TBCCC 2016), at the expense of less tolerant Redwood, Douglas Fir, and montane hardwood forests (Micheli et al. 2016). Ranges for wetland plants, as well as wetland plant community composition, may shift as these plants respond to rising temperatures and changes in water quality and availability (Short et al. 2016, Gillard et al. 2017). Changes in wildfire patterns can be expected to affect different plant communities in varying ways, depending on fire intensity and interval; short intervals between high-intensity fires can reduce regeneration in some plant communities (Diaz-Delgado et al. 2002, Moritz et al. 2011, Ferriter 2017). However, significant uncertainty in projected future rainfall yields subsequent uncertainty in the kinds of vegetation that will dominate the future landscape, and land management practices will also

Highway notice in Cotati. Photo: Tony Webster. CC by SA 2.0.

influence vegetation patterns (Chornesky et al. 2015). Furthermore, predicted vegetation patterns under climate change represent end states that may take decades or centuries to manifest (Ackerly et al. 2015).

Drought

Although future annual precipitation totals are uncertain, an increase in the length of the Laguna's dry season is likely (CH2M Hill 2015, Mount et al. 2017). Rising temperatures will increase the likelihood that low-precipitation years coincide with warmer summers, inducing drought (Diffenbaugh et al. 2015). While extreme heat events pose an immediate short-term threat to human health, extended warm droughts affect watershed functionality over long periods of time. For example, after the historic drought of 2012-2016, large rain events only modestly replenished reservoirs like Lakes Sonoma and Mendocino, while recharge and soil moisture within the Russian River watershed were so low that it was projected to take several years of normal precipitation conditions to recover (Flint et al. 2018). The 2012-2016 drought also resulted in massive forest dieoffs (Asner et al. 2016), and future droughts are expected to do the same (Das et al. 2013). The lingering effects of drought are especially acute in higher elevations with thinner soils and lower groundwater capacity. Further downstream, drought conditions may also disrupt wastewater treatment processes, which require certain amounts of water to function properly (Chappelle et al. 2019).

Wildfire

As summer conditions become hotter and drier, wildfires are likely to become more frequent and more destructive in the region, and across California (Fried et al. 2004, Krawchuk and Moritz 2012). Future wildfires are likely to endanger human lives, property, and wildlife (Krawchuk and Moritz 2012), and alter hydrologic and geomorphic processes that could result in further risks (Moody and Martin 2009, Coombs and Melack 2013). With respect to human impacts, there are risks associated with direct wildfire contact and contact with the airborne particulates wildfires produce (Tarnay 2018). However, human actions can largely determine the extent of wildfire damage as the climate changes (Mann et al. 2016). Land managers can reduce local fire risk by reducing fire ignitions, proactively managing land to prevent wildfires, and promoting land cover types more resistant to fire (e.g., shrublands and closed woodlands rather than invasive-dominated grasslands) (Keeley 2001, Bowman et al. 2011). Additionally, city planners can encourage higher density LID in more defensible urban centers, as opposed to expanding in the wildland-urban interface, where communities are more vulnerable.

Building Climate Resilience

As the populations of towns and cities surrounding the Laguna grow (SCEDB 2018), more and more people will witness these effects of climate change. Land use changes concomitant with increasing populations may exacerbate these effects by, for example, increasing urban runoff, creating urban heat islands, and increasing exposure to wildfire. However, through LID and urban greening, local municipalities can enhance their resilience to climate change while ensuring the ongoing health of the Laguna (Pyke et al.2011). Outside of urban centers, restoration activities and agricultural best management practices can mitigate the effects of climate change described above. Restored wetlands and riparian areas will slow and retain flood runoff, filter pollutants and nutrients, and recharge groundwater, allowing for more resilience to changing climate conditions (Seavy et al. 2009). Metrics such as the recently released Climate Resilience Outcome Metrics developed by the State of California (CNRA 2018) can be used to track the impact of management and restoration actions on landscape resilience over the long term. §

Native California trees in an urban settting. Photo: Shira Bezalel.

The Future of Flood Flows in the Laguna

Recently, the USGS and Sonoma Water conducted hydrologic modeling to assess the possible change in future Laguna flows at mid- and late-21st century compared to historical conditions, focusing on unimpaired flow (i.e., natural flow conditions without dams and major diversions). The analysis included the four downscaled global climate models (GCMs) considered by California's Climate Action Team to be a good representations of possible future conditions. HadGEM2-ES is the "warm/dry" conditions model, CNRM-CM5 is the "cool/wet" conditions model, CanESM2 is the "average" conditions model, and MIROC5 is unlike the other three and chosen for better coverage of different possibilities (Pierce et al. 2018). For CNRM-CM5, CanESM2, and HadGEM2-ES, the analysis shows an increase in both the average and maximum number of days per decade the Laguna flow reaches 6,000 cfs (a daily mean flow that occurred about one day per decade in the 20th century), with the maximum days per decade flows reach 6,000 cfs increasing considerably by the end of the 21st century.

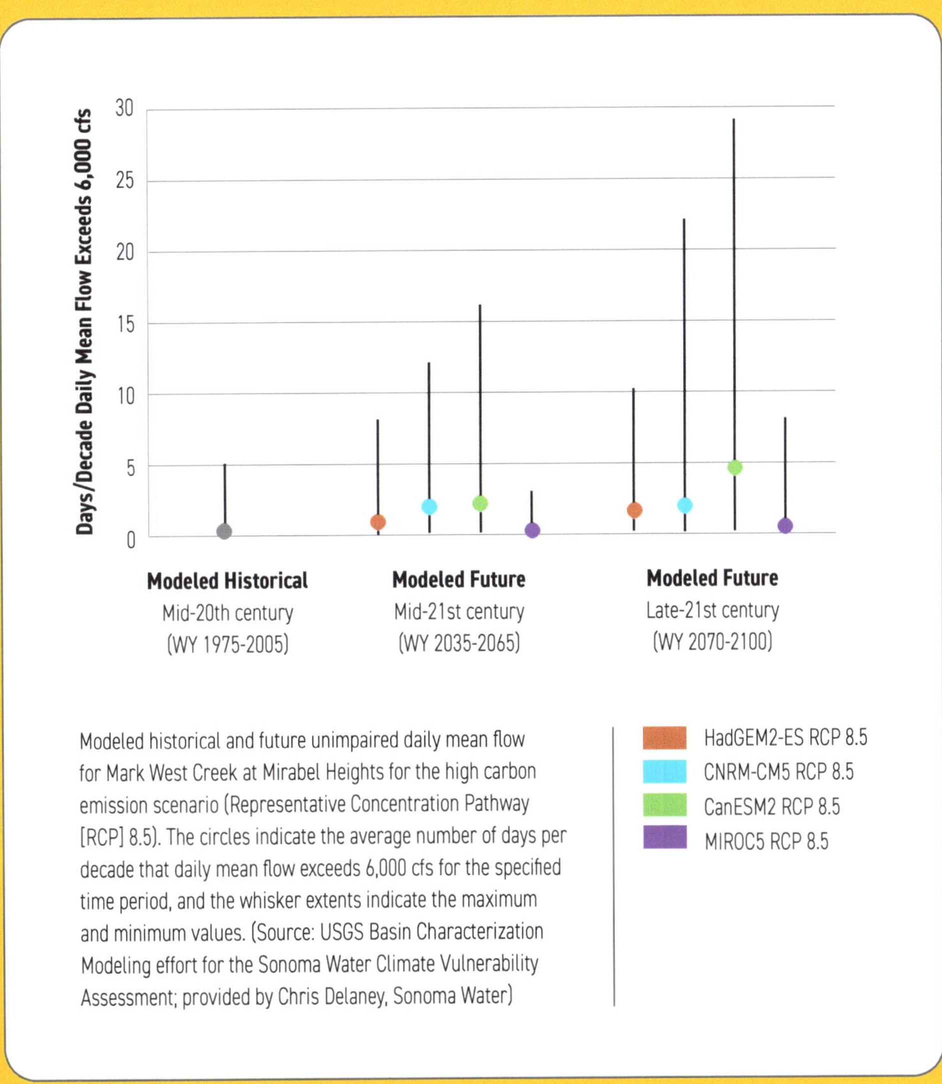

Modeled historical and future unimpaired daily mean flow for Mark West Creek at Mirabel Heights for the high carbon emission scenario (Representative Concentration Pathway [RCP] 8.5). The circles indicate the average number of days per decade that daily mean flow exceeds 6,000 cfs for the specified time period, and the whisker extents indicate the maximum and minimum values. (Source: USGS Basin Characterization Modeling effort for the Sonoma Water Climate Vulnerability Assessment; provided by Chris Delaney, Sonoma Water)

Fields and oak woodlands in the Laguna de Santa Rosa. Photo: SFEI.

7 Restoration Vision

The Laguna de Santa Rosa Vision map highlights the areas of greatest restoration opportunity within the Laguna study area, as identified through the visioning process. The areas displayed are intentionally aspirational, and it is hoped that they can serve as a spur to the imagination of a future landscape in which people and nature can both thrive. A range of actions is recommended, including habitat restoration, riparian enhancement, and channel recontouring. Additional changes to infrastructure that would improve ecological functioning within the Laguna in the future are also included.

Opportunity areas initially identified by the TAC to help achieve Vision goals and objectives were refined to enable quantification of the benefits of increases in habitat area that restoration could achieve. They were also refined to reflect basic physical controls and suitability for different wetland types based on elevation, soil type and drainage class, and depth to groundwater, and to exclude current wetland land covers (see Appendix C for details).

People will continue to be an integral part of the Laguna, which serves as a place where they live and work. Covering all areas of the Laguna study area with potential restoration opportunities is not the intent of the Vision. The Vision map focuses on wetland (as opposed to upland) restoration opportunities, because wetland and riparian habitats have been the most transformed over time. Note that it is not the intent of the Vision to preclude restoration opportunities outside the areas displayed, and restoration opportunities that arise outside the areas should also be considered.

The Laguna is part of its surrounding landscape, and it will be important to maintain a broad perspective, managing the watershed as a whole. Many of the recommended actions depicted here will depend on successful management of processes outside the study area. Expected benefits to landscape resilience as a result of restoration actions proposed within and outside the Laguna are on pages 110-121. Key recommendations for watershed-wide actions follow on page 122.

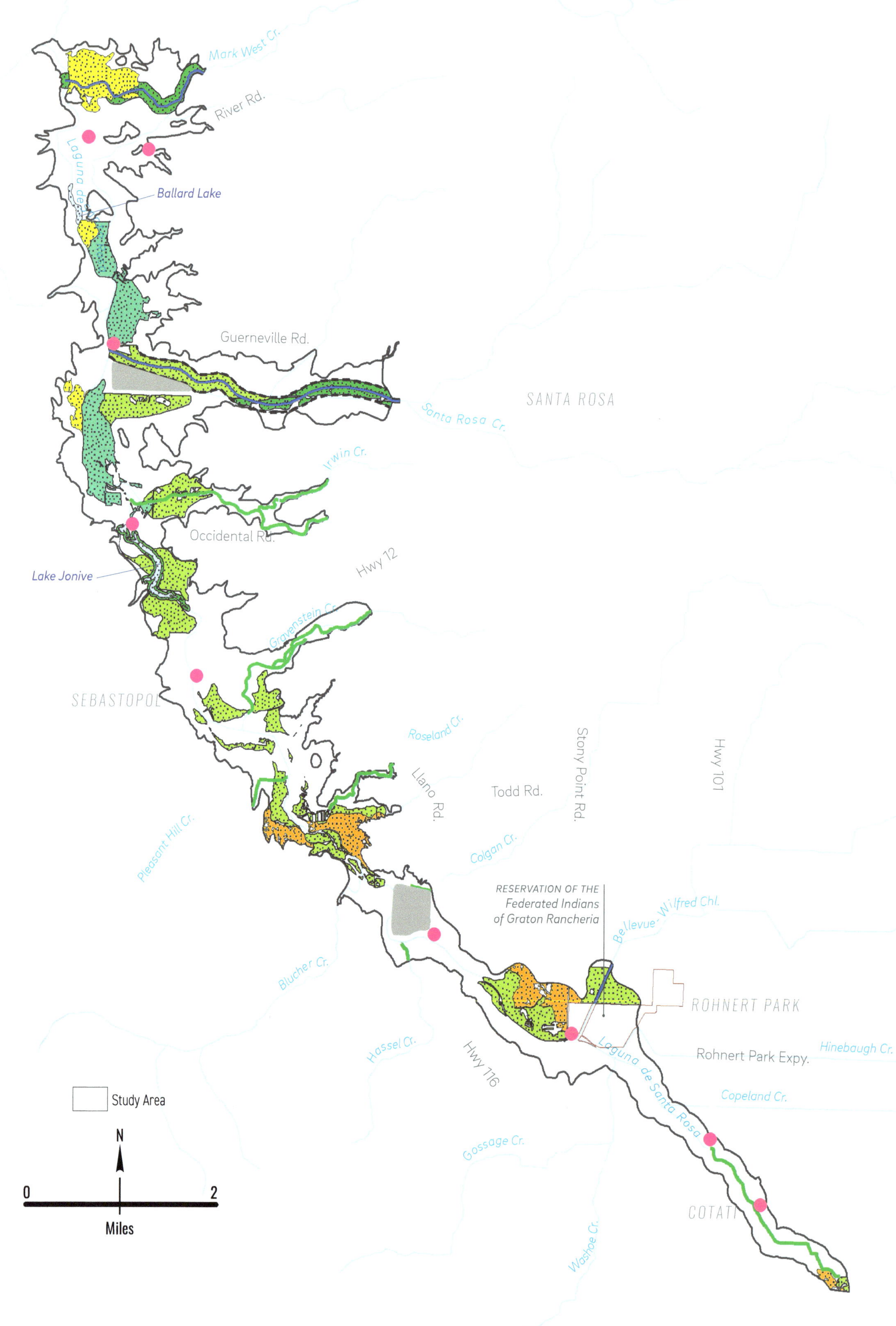

Mark West Cr.
River Rd.
Laguna de Santa Rosa
Ballard Lake
Guerneville Rd.
SANTA ROSA
Santa Rosa Cr.
Irwin Cr.
Occidental Rd.
Hwy 12
Lake Jonive
Gravenstein Cr.
SEBASTOPOL
Roseland Cr.
Llano Rd.
Todd Rd.
Stony Point Rd.
Hwy 101
Pleasant Hill Cr.
Colgan Cr.
RESERVATION OF THE
Federated Indians
of Graton Rancheria
Bellevue-Wilfred Chl.
Blucher Cr.
ROHNERT PARK
Hassel Cr.
Hwy 116
Laguna de Santa Rosa
Rohnert Park Expy.
Hinebaugh Cr.
Copeland Cr.
Study Area
N
Gossage Cr.
0
2
Miles
COTATI
Washoe Cr.

Laguna de Santa Rosa
Restoration Vision Map

- **Vision implementation would greatly expand wetlands,** with the aim of better providing functions such as life history support for amphibians, reptiles, fish, and birds; as well as nutrient and pollution regulation.

- **Widening and enhancing riparian zones through levee setbacks and reducing flow constrictions can reduce flood flows,** and provide enhanced habitats through increased shading and greater native vegetation cover.

- **Configuration of wetland and riparian areas would be different from historical,** but with careful placement, lost functions can be restored with thoughtful restoration design and implementation.

- **Enhancement and maintenance of habitats through planting of native species and management of weeds** will be essential to successful restoration.

- **Success of these restoration actions within the Laguna will in part depend upon management of key drivers** of flow, groundwater, sediment, and nutrients at a watershed scale

Wetland and Aquatic Restoration Opportunities

Open Water • Shallow and deep pools

Freshwater Marsh Complex • Freshwater marshes with diverse vegetation such as river bulrush and cattail.

Wet Meadow Complex • Diverse mix of wetland vegetation such as sedges and rushes.

Willow Forested Wetland Complex • Dense stands of willow vegetation intermixed with herbaceous species.

Mixed Riparian Forest • Woody riparian canopy trees such as box elder, willow, and oak, with shrub and herbaceous understory.

Oak Savanna and Vernal Pool Complex • Seasonal wetland pools within a matrix of oak savanna and oak woodland.

Riparian Management Opportunities

Channel and Levee Realignment • Infrastructure modifications to increase conveyance and riparian habitats.

Riparian Enhancement • Vegetation management within riparian areas to favor native species.

Additional Management Actions

Bridge Crossings • Redesign bridges when ready for replacement.

Wastewater Treatment Infrastructure • Explore options for relocation in the long-term.

Vision Concepts

The Vision actions constitute a blend of short-term and long-term concepts that can help restore lost habitat, create new beneficial habitat, or represent changes to infrastructure that support ecosystem processes. Realization of any of these concepts will require partnerships and cooperation with willing landowners, and careful consideration of cost and feasibility of implementation. Sequencing of actions will also require consideration, since conditions at another site may affect function and success of restoration actions at a project site. This section provides details for the concepts presented in the Vision map.

Recommendations for restoration cover six broad types of habitat, ranging from open water to oak woodland and savanna. Though the Vision depicts large, contiguous patches of restored habitat types with discrete boundaries, the scale and simplicity of the map belies the actual complexity of local conditions. Wetlands in the Laguna are complex and heterogeneous, with a complex mix of topography and species composition, depending on local conditions. For example, the Vision map depicts discrete boundaries between wetland types, while in reality, wetland systems often include core areas of wetland habitat with ecotones that transition from wetter to drier. As such, a single wetland restoration project may encompass areas that transition between open water, freshwater marsh, riparian, wet meadow, and/or oak savanna, depending on its scale and landscape position. Varied wetland elevations and vegetation types provide beneficial variations in habitats. For example, a greater diversity of water depths in wetlands is likely to support a greater diversity of shorebirds and waterfowl that use different water depths (Isola et al. 2000). The appropriate vegetation cover and wetland type will depend on specific site conditions and restoration goals. Successful restoration along the Laguna Middle Reach has incorporated many types of land cover, and can serve as a demonstration for potential future restoration areas.

Promoting habitat complexity and heterogeneity is an important part of habitat restoration. Successful habitat restoration should take the variability of local conditions into account, such as elevation, soil type, and hydrologic regime when making restoration designs, as well as during monitoring and maintenance.

Wetland and Aquatic Restoration: **OPEN WATER**

Pools and lake-like features historically provided cool, open water areas that likely provided valuable habitat for juvenile coho salmon, other salmonids, and native fishes, as well as waterfowl. Remaining areas of summertime open water are now reduced in total extent, are shallower due to increased sedimentation, and in many cases, are covered with undesired aquatic weeds. Despite this, coho salmon persist in the Laguna and in Mark West Creek (Merritt Smith 1995; M. Obedzinski, pers. comm), and restoring favorable pool habitats may benefit coho and other native fishes.

Action: Re-create or enhance deep open water areas by dredging to reset sediment accumulation. Plant native vegetation to shade the edges of open water, and manage water quality within pools. Remove and manage invasive species like Ludwigia hexapetala.

Benefits: Increased foraging, breeding, and resting habitat for waterfowl, wading birds, and fish. Planting of native trees reduces water temperatures and provides habitat for resident and migratory birds. Dredging and removal of nutrient-laden sediments can reduce nutrient concentrations. Improved conditions for recreation on Lake Jonive.

Locations: Lake Jonive, restored Ballard Lake.

Considerations: The long-term viability of deep, cool pools in the Laguna could be compromised by continued deposition of excess fine sediment, excess nutrients, and projected rises in air and water temperatures, all of which affect water quality within pools. Landscape-scale sedimentation sources will need to be reduced in order to prevent the need for ongoing maintenance dredging of deep pools. Water quality monitoring and management within pools will be needed to ensure high quality aquatic habitat for native fishes.

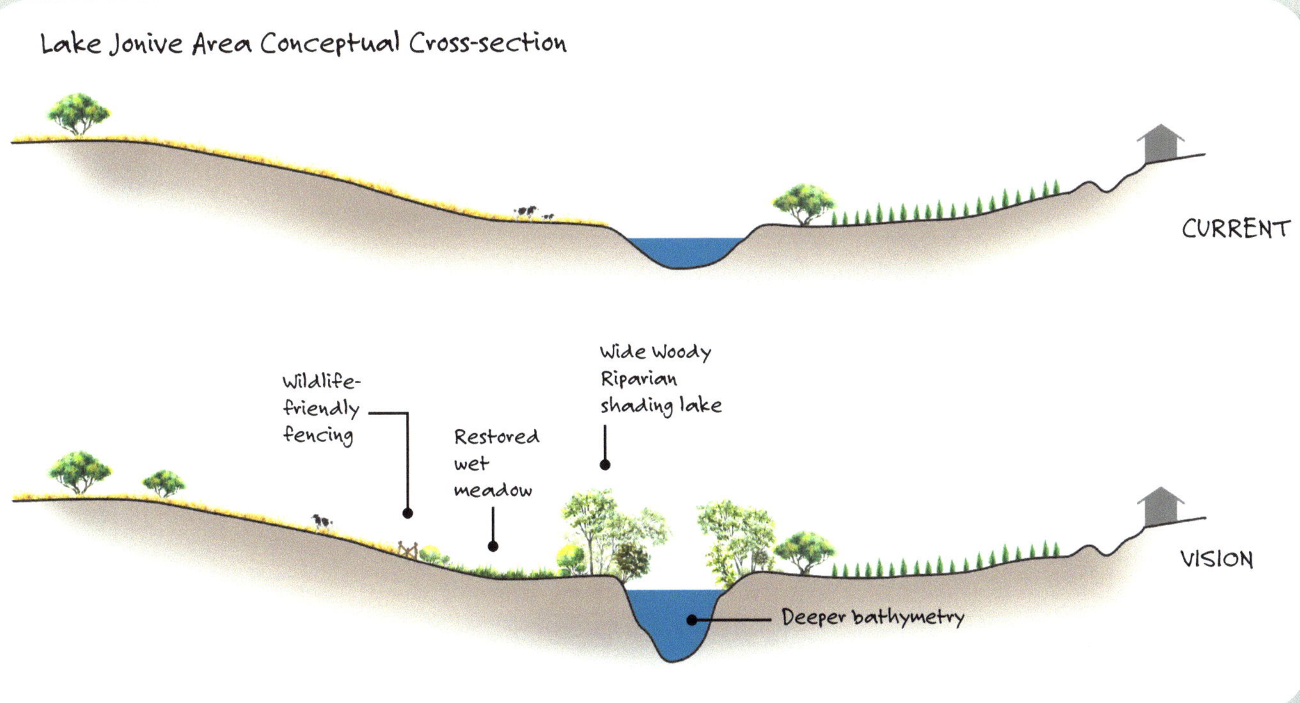

Wetland and Aquatic Restoration: FRESHWATER MARSH COMPLEX

Freshwater marshes were historically most prevalent in the northern, downstream reaches of the Laguna, and have been widely converted to seasonal agricultural uses. Increased duration and area of flooding has in some cases made planting crops within these areas infeasible. In areas with appropriate hydrology and groundwater conditions, it is possible to restore freshwater marshes that will provide valuable habitat as well as filter nutrients and pollutants before they reach the Laguna.

Action: Create freshwater marshes with diverse topography and vegetation types appropriate to site conditions. Experimentally test different plant communities for their ability to respond to invasive species pressure. Manage invasive species.

Benefits: Sediment trapping, nutrient assimilation and processing, carbon sequestration; nesting, foraging and resting habitat for marsh wildlife in summer, and for fish when marshes are submerged in winter.

Locations: Greatest opportunities lie north of Occidental Road.

Considerations: Freshwater marshes require the proper hydrologic conditions, with access to surface and groundwater that support them. Care needs to be taken in designing potential marshes to provide topographical variability and transition zones both to wetter and drier cover types. When maintaining marshes, ensure the areas host a desirable assemblage of native plant species, such as river bulrush, cattail, and Santa Barbara sedge, rather than undesirable invasive plant species.

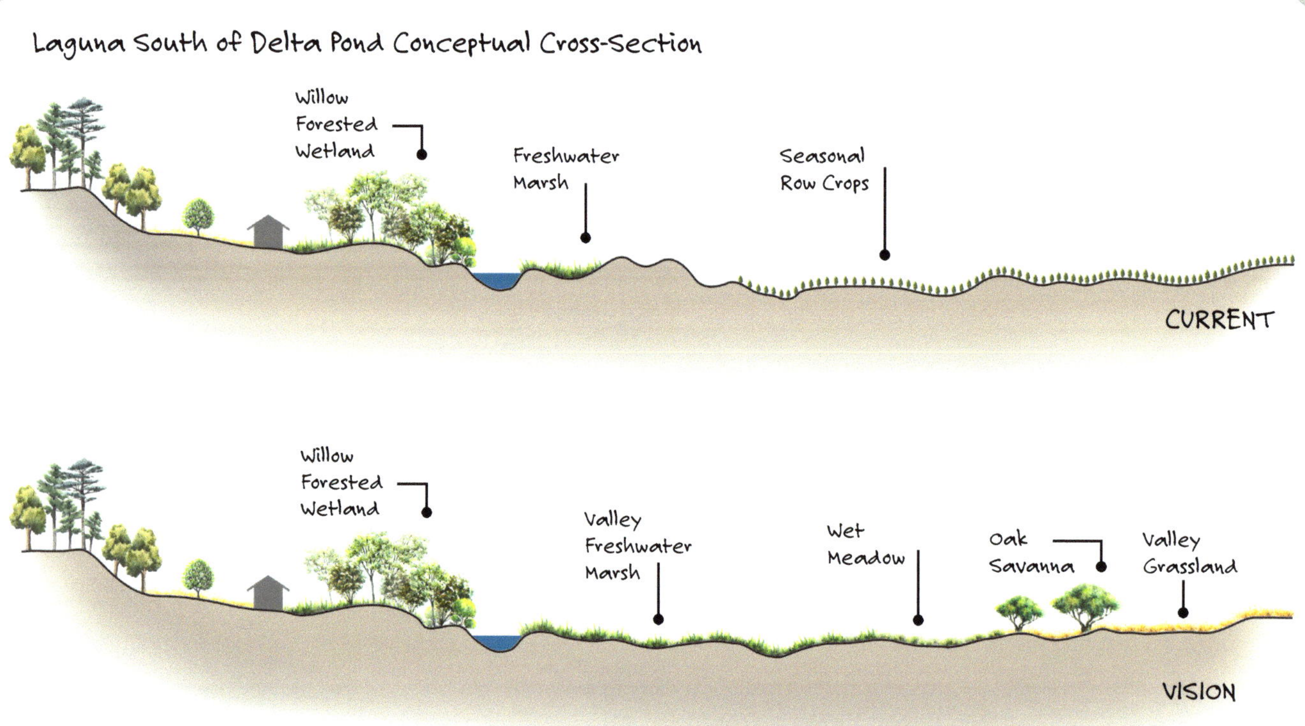

Wetland and Aquatic Restoration: WET MEADOW COMPLEX

Expanding and creating wet meadow complexes presents the greatest potential opportunity for habitat restoration in the Laguna, in part because this cover type is the most lost compared to historical conditions. Wet meadows host a unique variety of vegetation and provide habitat for a wide range of wildlife, including nesting and foraging habitat for amphibians, birds and fish. Their nutrient filtration and processing function varies by plant assemblage and landscape position, and it would be possible to learn about optimal landscape arrangement and species composition through experimental plantings within restoration areas. There is potential to work with Native People to create educational workshops in traditional uses of wet meadow plants to strengthen cultural connection to these lands.

Action: Restore wet meadow complexes adjacent to wetter habitats, and include transitions to drier habitats; test nutrient and sediment capture properties of different plant assemblages.

Benefits: Increased plant biodiversity, productive floodplain habitat for fish, nutrient and sediment trapping and assimilation, opportunity for enhancing cultural resources, e.g., by planting and harvesting culturally important species such as basket sedges.

Locations: Laguna between River and Guerneville Roads, south of Delta Pond and Highway 12, Laguna Middle Reach including the CDFW Cooper Road Unit area south of Lake Jonive, area north of Bellevue-Wilfred Channel.

Considerations: Many of the same considerations for freshwater marshes apply also to wet meadows, including the need for access to surface and groundwater, topographical variability, transition zones both to wetter and drier cover types, planting and maintenance of native species, and invasive species control.

Santa Barbara sedge (*Carex barbarae*) among alders near Irwin Creek. Photo: SFEI.

Wetland and Aquatic Restoration: WILLOW FORESTED WETLAND COMPLEX

Historically, willow forested wetlands occupied much of the area near channels in the north and west portions of the Laguna. Their complex mix of vegetation, including willow trees, Oregon ash, tules, and sedges, hosted a wide variety of wildlife. Generally, these wetlands do not have well-defined channels and tend to be semi-permanently flooded with shallow water. This type grades with drier riparian forest types, and wetter freshwater marsh types. Restoring this cover type can help provide cool water and food for native fishes.

Action: Restore or enhance diverse willow stands, especially in the northwestern area of the Laguna where historical stands were once present.

Benefits: Wooded, nearly-permanent flooded areas are beneficial to nesting and foraging songbirds and provide floodplain-derived food for fish, provide shading of creek channels that lowers water temperatures, and provide sediment and nutrient trapping to improve water quality.

Locations: Northwest region of the study area, Laguna south of Santa Rosa and Irwin Creek confluences.

Considerations: Late-season flooding in areas identified on the Vision map may hamper vigorous growth and development of this vegetation type. Study of the hydrologic regime should be undertaken as part of restoration designs in these areas.

Willows (*Salix* spp.) along lower Santa Rosa Creek. Photo: SFEI.

Wetland and Aquatic Restoration: OAK SAVANNA and VERNAL POOL COMPLEX

Historically, this habitat complex occurred through much of the Santa Rosa Plain. This habitat complex hosts a unique assemblage of plants and animals adapted to its seasonal pool formation. Ensuring that oak savanna and vernal pool areas are well-connected to adjacent wetlands and uplands can provide high-quality habitats for birds and reptiles. Though most oak savanna and vernal pool complexes in the Laguna watershed occur outside the study area, there are a few areas of these complexes that can be protected, enhanced, or managed within the study area.

Action: Protect, enhance, and restore vernal pool complexes.

Benefits: Increased populations of rare and endemic vernal pool plants and animals.

Locations: Focus on areas south of Roseland Creek outside the 10-year floodplain.

Considerations: Recommendations on the Vision map focus on preservation and restoration in areas that are seldom inundated by river flooding (i.e., that are outside the estimated 10-year flood area), since connection to riverine flooding exposes rare vernal pool animals to predation. Preservation of existing vernal pool complexes benefits vernal pool species more than creation of new pools for mitigation. There are many vernal pool preservation and enhancement opportunities on the Santa Rosa Plain outside of the study area.

Oak Savanna / Vernal Pool Complex. Photo: SFEI.

Wetland and Aquatic Restoration: **MIXED RIPARIAN VEGETATION MANAGEMENT**

In addition to the expansion of riparian habitats, riparian vegetation enhancement is an important part of restoration. Many areas along the Laguna and its tributaries now host non-native species that reduce aesthetic, habitat and water conveyance values, and management that favors native species can address these issues. Appropriate vegetation in enhancement areas ranges from herbs and grasses to shrubs and trees. Though not historically present, planting trees along Laguna tributary channels can reduce water and air temperatures. The urban area around Cotati and Rohnert Park is highlighted for riparian enhancements in the Vision map; but many projects undertaken in the Laguna will have a riparian component, and riparian vegetation management that emphasizes native species will apply to all of them.

Action: Increase native tree, shrub, and/or herbaceous vegetation cover in riparian areas. This includes: vegetation management in public rights-of-way, e.g., through Sonoma Water's Stream Maintenance Program; in urbanized areas where buffers are narrow; as well as in wide riparian buffers located in private and public rural lands. Riparian enhancements should be site-specific, and should include areas of herbaceous plants alongside channels where appropriate.

Benefits: Native plants benefit from removal of invasive species and animals benefit from increased native plant cover, which provides habitat. Healthy riparian vegetation provides filtration of fine sediments and pollutants that can improve water quality, and reduces water and air temperatures through shading. Well-managed plants provide aesthetic value.

Considerations: Vegetation within flood control channels can impede flows, so selection of vegetation compatible with this function will be needed. Control of non-native species in riparian zones will be a long-term commitment.

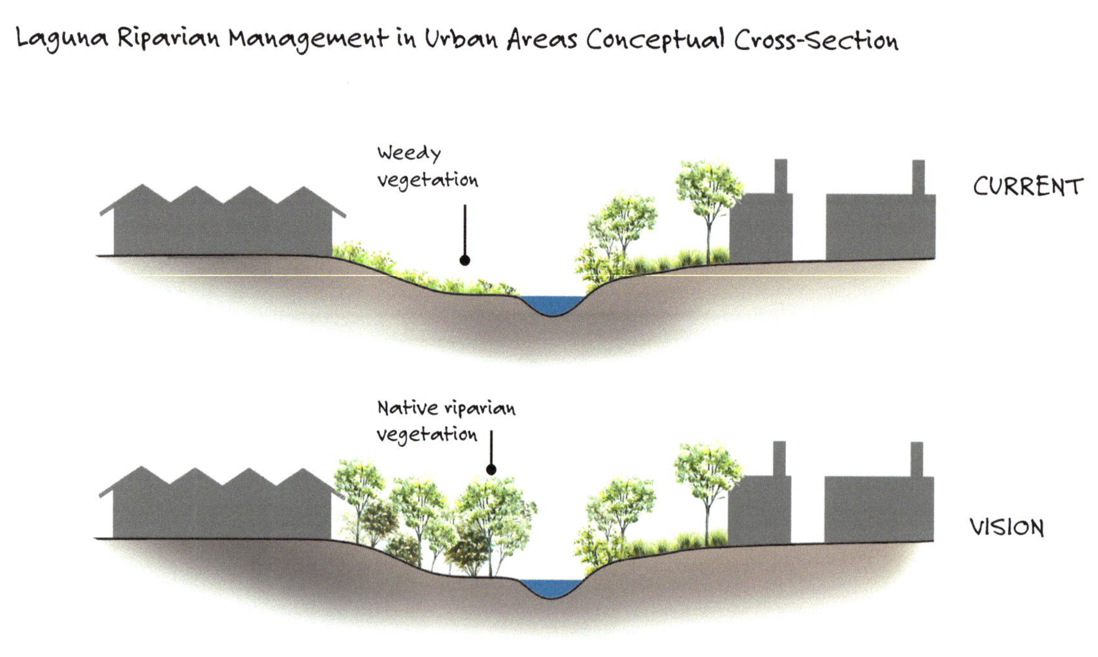

Mixed riparian vegetation along Copleand Creek. Photo: SFEI.

Box elder (*Acer negundo*) leaves. Photo: SFEI.

Riparian Restoration: **MARK WEST CREEK**

Over time, the confluence of Mark West with the Laguna has been realigned to maximize farmable land, changing the delivery of sediment to the Laguna, and contributing to the historical filling of Ballard Lake in the 1930s and 1940s. Realigning the creek could improve habitat for migrating salmon.

Action: A channel realignment, or the addition of a bypass channel connection near the historical confluence with the Laguna. If channel realignment were undertaken, restoration of a large area of willow forested or other wetland complex near the confluence with the Laguna is recommended. Upstream of the potential new confluence, a wide mixed riparian buffer is recommended.

Benefits: Salmonids may benefit from an improved migration route. Restored willow forested wetland and riparian vegetation would provide shading for cooler water temperatures, filtration of pollutants, and floodplain habitat, providing food and cover for resident and migratory fish, birds and mammals.

Considerations: The alignment shown is conceptual and matches the historical Mark West channel (ca. 1850s). Actual realignment alternatives would need to include careful study of predicted flow patterns and monitoring for salmonid stranding during drawdown of Laguna floodwaters, as well as effects on land uses and transportation infrastructure.

Lower Mark West Creek after a winter rain. Photo: Steve Shupe. Creative Commons.

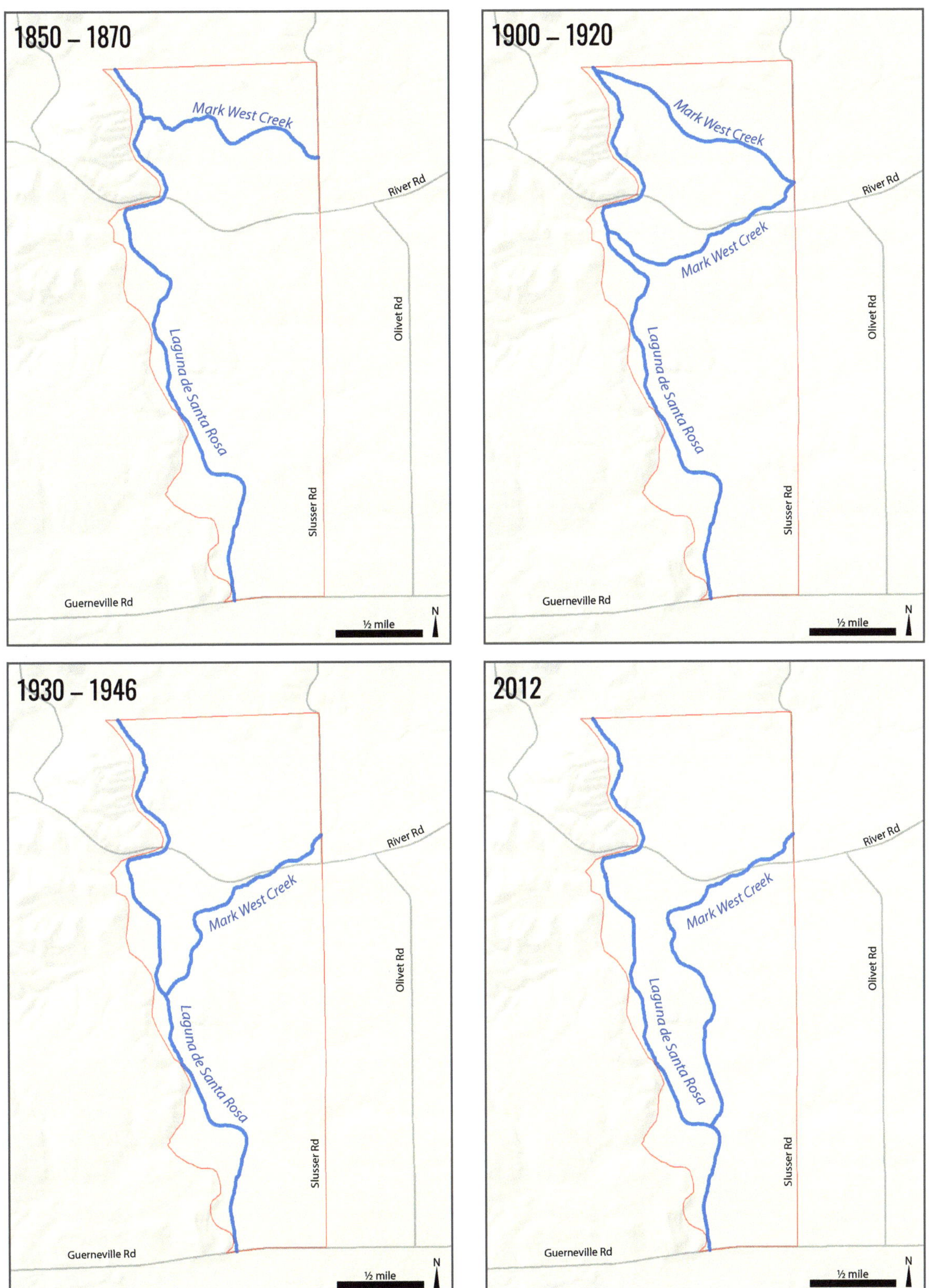

Alignments of Mark West Creek over time. The confluence of Mark West Creek has been moved approximately 3.2 km (2 mi) from its location in 1850 to control flooding and shift sedimentation patterns. Source: Baumgarten et al. 2014.

Riparian Restoration: SANTA ROSA CREEK

Formerly a wide, braided, multi-channel system with a rich mosaic of grassland, woody riparian, and wetland complexes, the lower reaches of Santa Rosa creek have been straightened and are now leveed. This straight alignment shunts water, sediment, trash, and debris quickly into the Laguna, exacerbating flow constriction at the confluence.

Action: Set back levees and realign lower Santa Rosa Creek. Enhance riparian habitats within the setback.

Benefits: Together with water and sediment control in the upper watershed, realigning the creek and setting back the levees could allow storm water energy to dissipate upstream of the Laguna, thus reducing sediment delivery. If there is sufficient space between the new levees, a braided channel system could form. Restoring native vegetation types and hydrologic processes within such a setback can provide essential services such as water temperature regulation, food production and physical habitat for wildlife, and sediment retention. Recreational opportunities are possible within the levees for community use when the river is not in flood stages.

Considerations: The realignment shown on the Vision map represents a conceptual alignment based on the average position of the main historical channel and a suggested setback width. Actual realignment and levee redesign would require specific site studies. Potential re-design of Delta Pond levees (considered below) would affect levee redesign for Lower Santa Rosa Creek.

Lower Santa Rosa Creek Conceptual Cross-Section

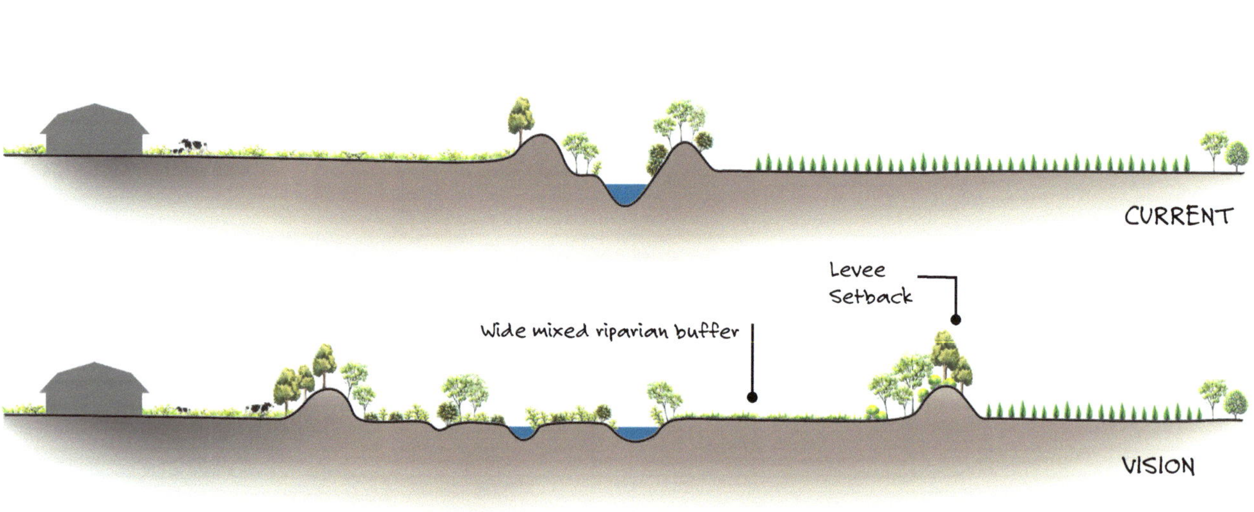

CURRENT

Levee Setback

Wide mixed riparian buffer

VISION

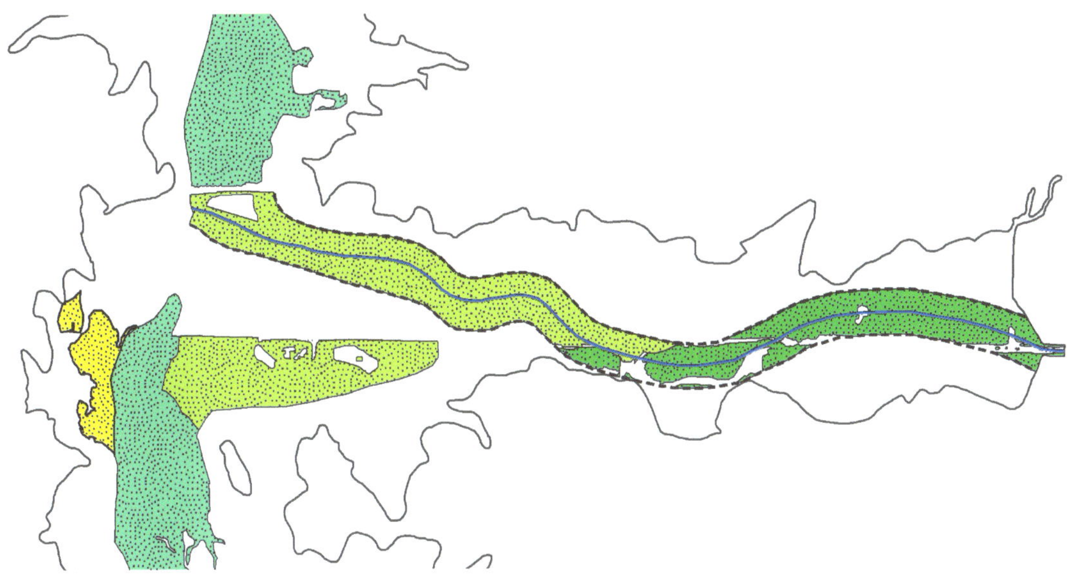

Conceptual alignment of Santa Rosa Creek, featuring a wide riparian buffer and wet meadow complex within a conceptual set back levee alignment.

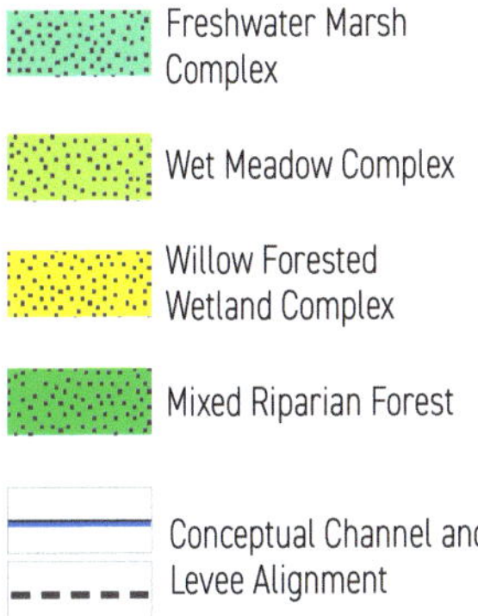

Lower Santa Rosa Creek, spring 2020. Photo: Wendy Trowbridge.

Additional Opportunities: BRIDGE CROSSINGS

Bridges can constrict channels and present barriers to wildlife movement, preventing animals such as bobcats, foxes, and deer from moving along the river corridor. Bridges with wider spans can alleviate channel constrictions and allow water, sediment, and wildlife to pass relatively unimpeded. As bridge infrastructure ages and replacement designs are contemplated, there is an opportunity to redesign them to better accommodate high flows and improve wildlife connectivity.

Action: When bridges are ready for replacement, re-design them with larger spans to convey greater flows and allow more room for wildlife.

Benefits: Reduced flow constriction, reduced sediment build up, greater habitat connectivity.

Locations: Prioritize Guerneville and Occidental Road Bridges.

Considerations: Infrastructure other than bridges, such as levees and berms, often contribute to local channel constrictions, and should be considered in conjunction with potential bridge redesigns.

Aerial view of the Guerneville Bridge. Source: NAIP 2016.

Additional Opportunities: LONG-TERM REDESIGN OF WASTEWATER TREATMENT INFRASTRUCTURE

The Delta Pond and the Laguna Wastewater Treatment Plant (WWTP) Ponds are key parts of regional municipal wastewater treatment infrastructure in the Santa Rosa area, holding treated water until it can be dispersed as irrigation in surrounding agricultural fields. In very wet years, water is sometimes released from these facilities into the Laguna. However, the levees forming this infrastructure constrict flow in the Laguna, which, together with increased sedimentation and flow from the watershed, contribute to increased late-spring flooding. In the case of the Laguna near Delta Pond, levees around the Gallo Wetlands and the Guerneville Road Bridge also contribute to channel constriction.

Action: Reconsider designs and location of Delta Pond, Llano Road WWTP Ponds, bridges, and nearby levees to reduce channel constriction and allow more space for the Laguna floodplain.

Benefits: Reduction of channel constrictions in the Laguna, reduced flooding damage, greater habitat connectivity.

Locations: Delta Pond near the confluence of Santa Rosa Creek, along with nearby levees; the Laguna WWTP near the confluence with Colgan Creek.

Considerations: The Delta Pond is located at the confluence of Santa Rosa Creek, which contributes much sediment to the Laguna. Sediment carried into the Laguna from Mark West Creek also contributes to flooding in the Laguna near Delta Pond. Designs should consider coordinated changes to infrastructure in this area, including levees and bridges. Essential functions that these treatment ponds provide for municipal wastewater treatment will still need to be fulfilled.

The Delta Pond during fall, when water levels are low. Photo: SFEI.

Benefits of Implementing the Landscape Vision

Implementation of the restoration concepts identified in the Vision would transform the Laguna landscape through changes in land cover type and configuration. Restoration of large patches of wetlands in thoughtful configurations can support ecosystem services such as life history support for amphibians, reptiles, fish, and birds; as well as aid nutrient and pollution regulation, flood management, and increase recreational and aesthetic values.

Though the Vision map depicts single wetland types as occupying large areas, it is important to note that actual wetland restoration projects would result in complex wetlands of varying types at the local scale, depending on site characteristics. Management to improve the quality of restored and existing habitats, including vegetation management to control weeds and promote wetland plant and animal diversity, is recommended.

The landscape presented in the Vision would not be a return to historical conditions, since that is neither possible, nor entirely desirable. However, by increasing the total area of wetland and riparian habitats, and reconfiguring habitats and infrastructure, important ecosystem services can be restored in a future Laguna landscape, with thoughtful restoration design and implementation. Quantitative changes in extent of habitat types, and metrics of landscape configuration that could be achieved through implementation of the Vision, are explored below.

Oaks in the flooded Laguna. Photo: SFEI.

HABITAT EXTENT

Significant increases in wetland land cover types.

Land cover changes in the Vision shift parts of the landscape from drier to wetter, by increasing the cover of wetland and riparian land cover types. Under the Vision, open water and pool areas would increase by nearly 20%, and mixed riparian vegetation would increase by over 50% compared to the current landscape. The biggest changes from existing habitat area would be large increases in valley freshwater marsh and wet meadow, which would both roughly double in extent, mainly due to conversion from farmed wetlands and oak savanna. Restoring wetland and riparian cover types would benefit a wide range of species, including marsh songbirds, raptors, reptiles, amphibians, and fish; and could enhance ecosystem services such as capture of fine sediments, nutrient processing, and flood attenuation.

HISTORICAL, MODERN, and VISION WETLAND COVER

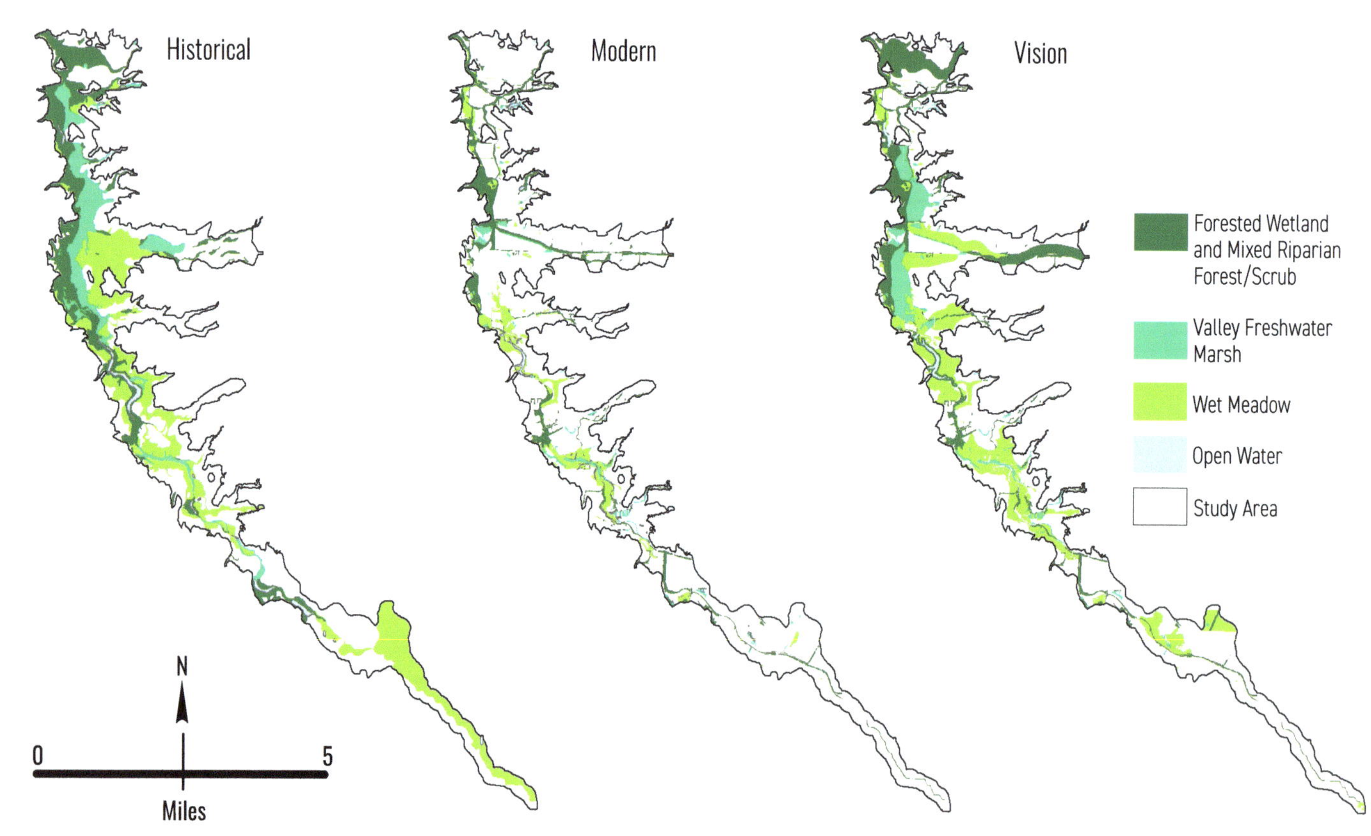

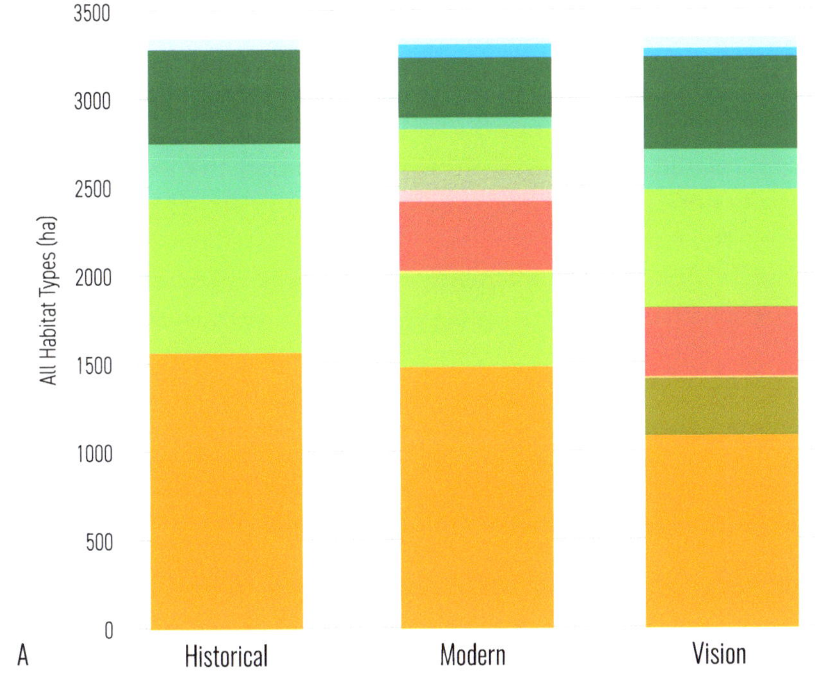

A. Change in habitat extent.
B. Change in wetland and aquatic habitat extent.

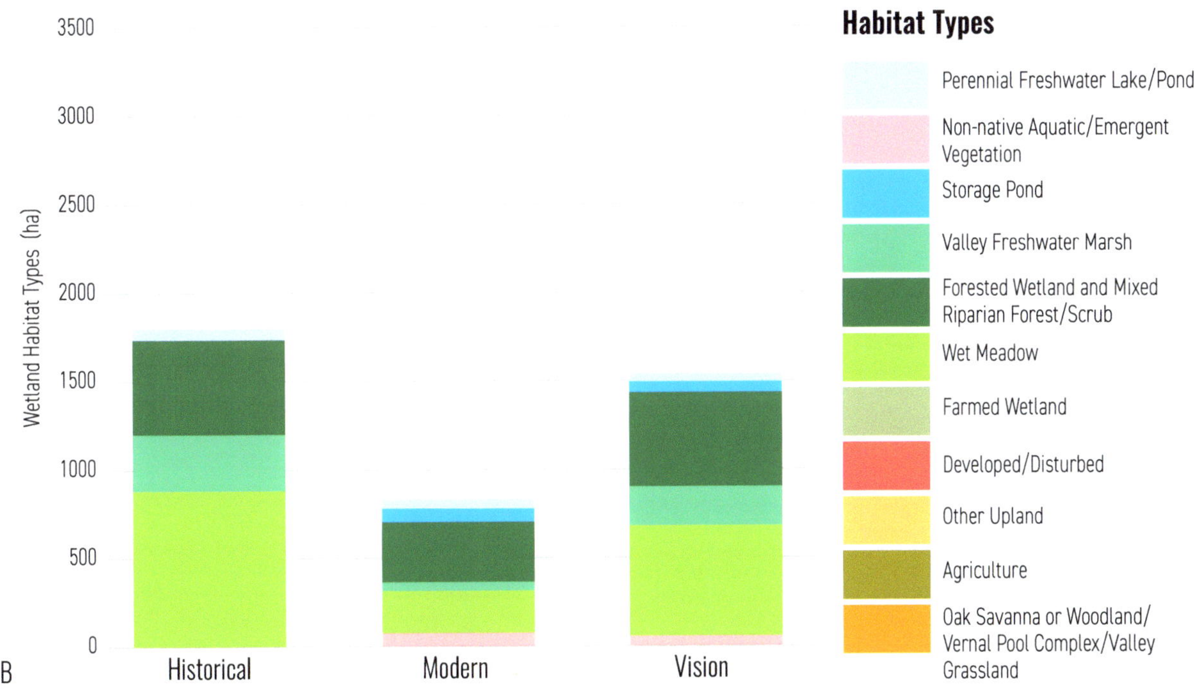

HABITAT CONFIGURATION

Larger, more connected freshwater marshes and wet meadows.

Implementing the Vision would lead to an overall increase in the extent of large wetland patches (i.e., discrete areas between 10-500 ha, 25-1,236 ac) for both freshwater marsh and wet meadow habitats. Large wetland patches support a wide variety of wildlife and produce valuable food resources for migratory fish and waterfowl. In addition, the Vision would result in increased connectivity between small and medium wetland patches with large patches, significantly increasing the total area of small and medium marsh patches within 500 m (1,640 ft) of a large patch. Increasing connectivity to large marsh patches provides a range of ecosystem benefits, including the support of local wildlife movement and dispersal.

Freshwater marsh plants submerged during winter. Photo: SFEI.

(Facing Page)

A. Freshwater marsh extent: historical on left, modern in middle, and future Vision on right.

B. Proportion of large, medium, and small wet meadow patches for historical, modern, and future Vision landscapes.

C. Marsh area by distance to nearest large patch.

HISTORICAL, MODERN, and VISION FRESHWATER MARSH

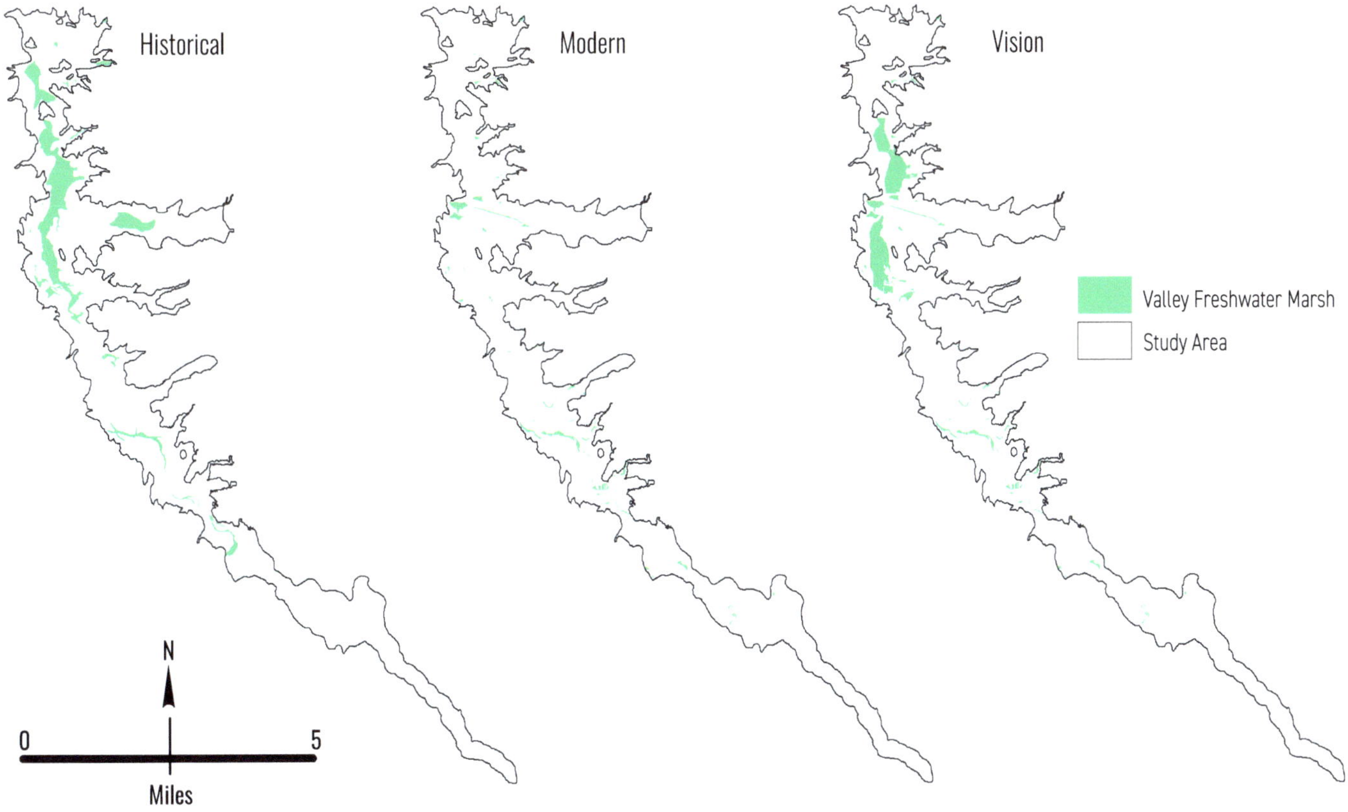

Historical

Modern

Vision

Valley Freshwater Marsh

Study Area

N

0 5

Miles

A

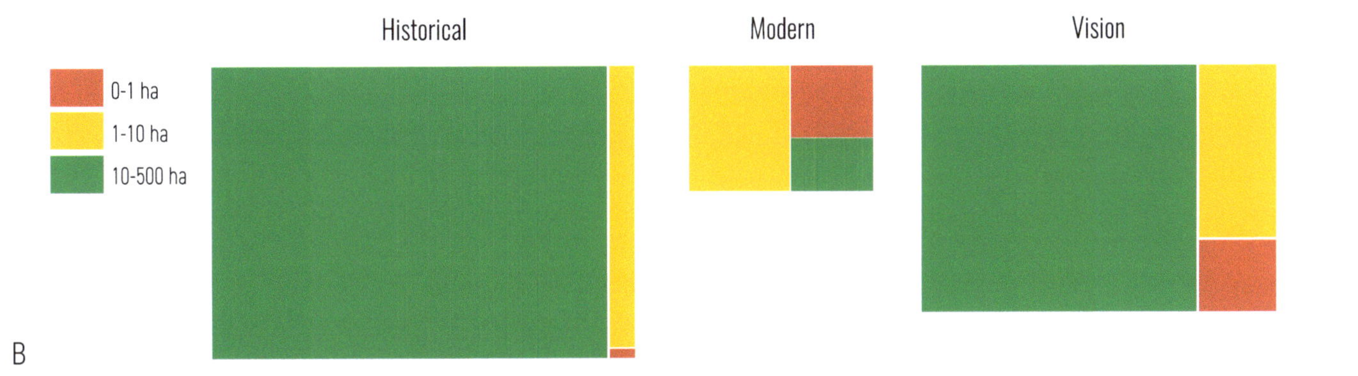

B

Distance to nearest large patch (m)	Area of freshwater marsh habitat (ha)		
	Historical	*Modern*	*Vision*
0-500	303	18	189
500-1000	1	2	1
1000-10000	12	34	33
>10000	0	6	2
Total	**316**	**61**	**225**

C

Planting *Carex barbarae*. Photo: Laguna de Santa Rosa Foundation.

(Facing Page)

A. Wet meadow extent: historical on left, modern in middle, and future Vision on right.

B. Proportion of large, medium, and small wet meadow patches for historical, modern, and future Vision landscapes.

C. Wet meadow area by distance to nearest large patch.

HISTORICAL, MODERN, and VISION WET MEADOW

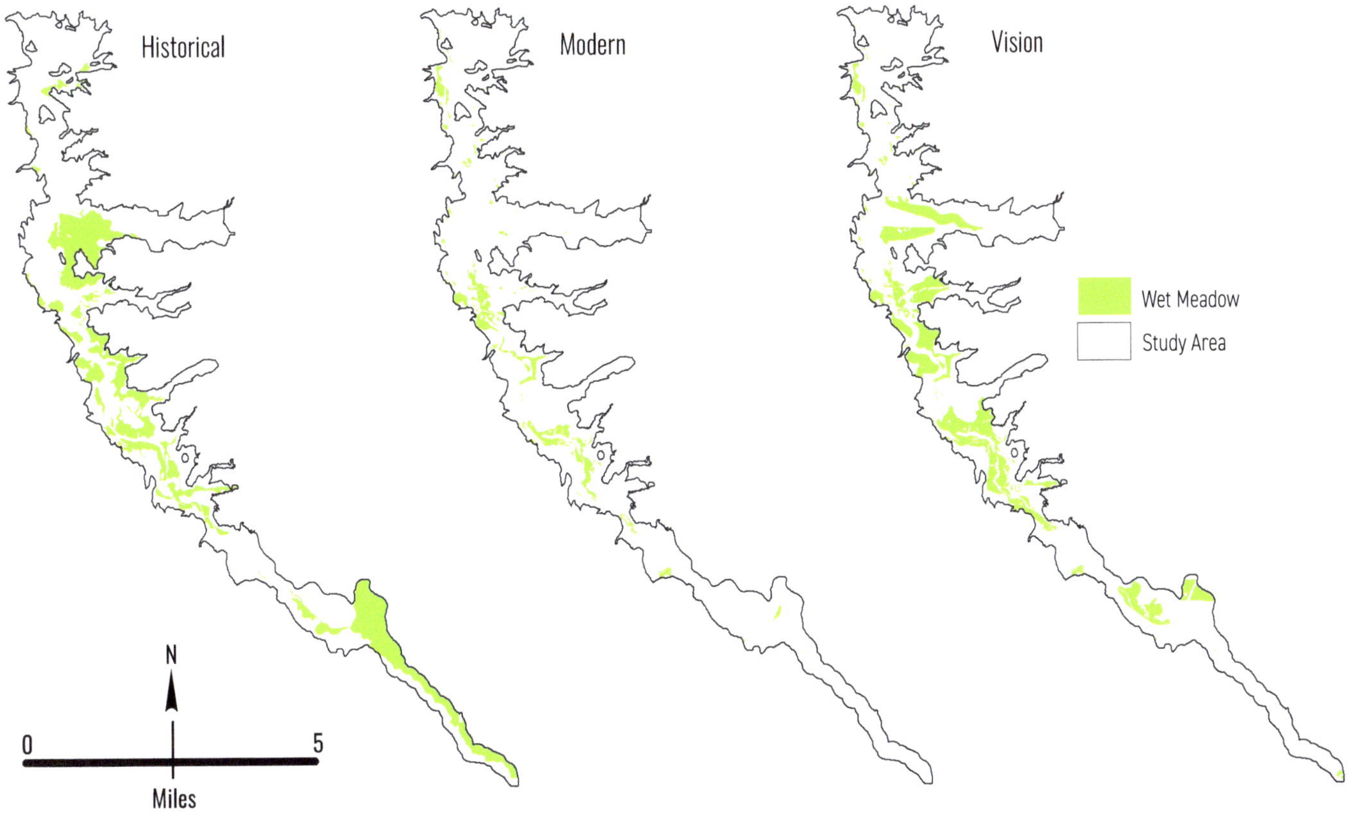

Historical

Modern

Vision

Wet Meadow

Study Area

N

0 5

Miles

A

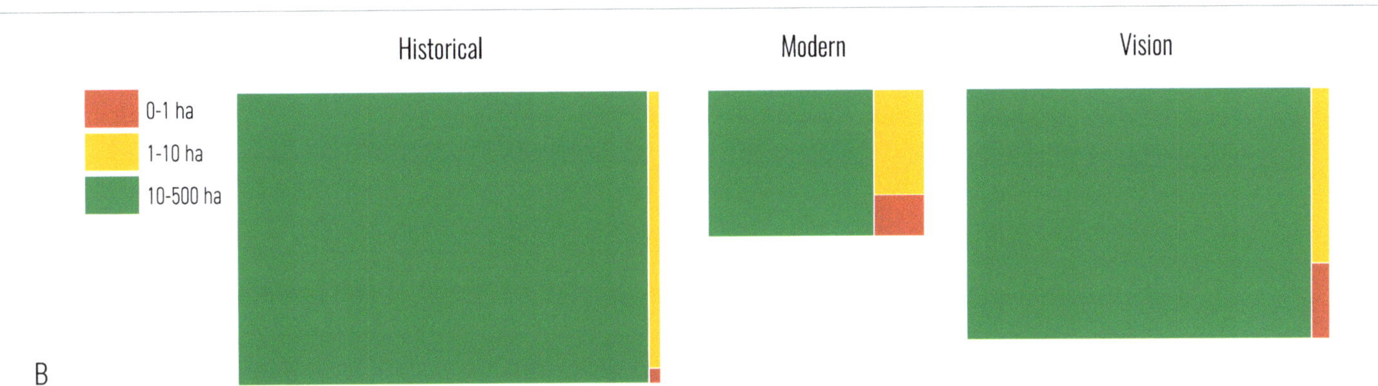

B

Distance to nearest large patch (m)	Area of wet meadow habitat (ha)		
	Historical	*Modern*	*Vision*
0-500	878	194	650
500-1000	6	16	13
1000-10000	5	23	6
>10000	0	0	0
Total	**888**	**234**	**734**

C

HABITAT CONFIGURATION

Increased riparian buffer width and connectivity.

Under the Vision, the total channel length with riparian forest would increase by ~10% (56 to 62 km, 35 to 39 mi) and the proportion of the channel network with wide riparian areas (i.e., areas with a width greater than 100 m, 328 ft) would increase threefold, from 10% to 30% of the total channel length. The extent of large riparian patches would increase, as would connectivity between small and medium riparian patches with large patches. As with freshwater marsh and wet meadow, the configuration of riparian patches would be different from historical, but Vision implementation would restore riparian connectedness and patch size to near historical levels. Improvements to riparian habitats would increase channel shading, increase native biodiversity, and increase filtering of pollutants.

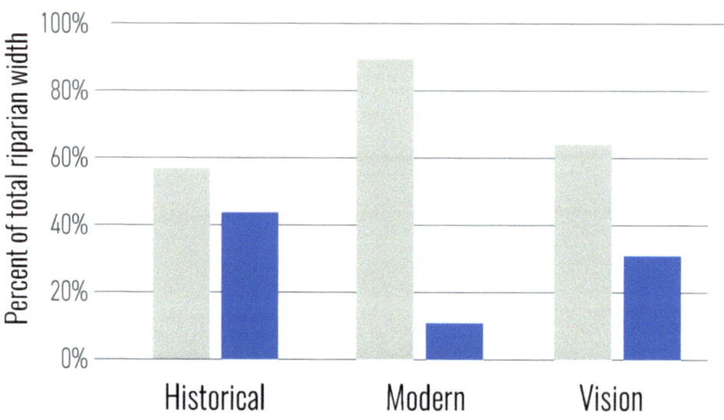

Proportion of total channel length with riparian patch width greater than, and less than 100 m.

(Facing Page)

A. Riparian extent: historical on left, modern in middle, and future Vision on right.

B. Proportion of large, medium, and small riparian patches for historical, modern, and Vision.

C. Riparian area by distance to nearest large patch.

HISTORICAL, MODERN, and VISION RIPARIAN

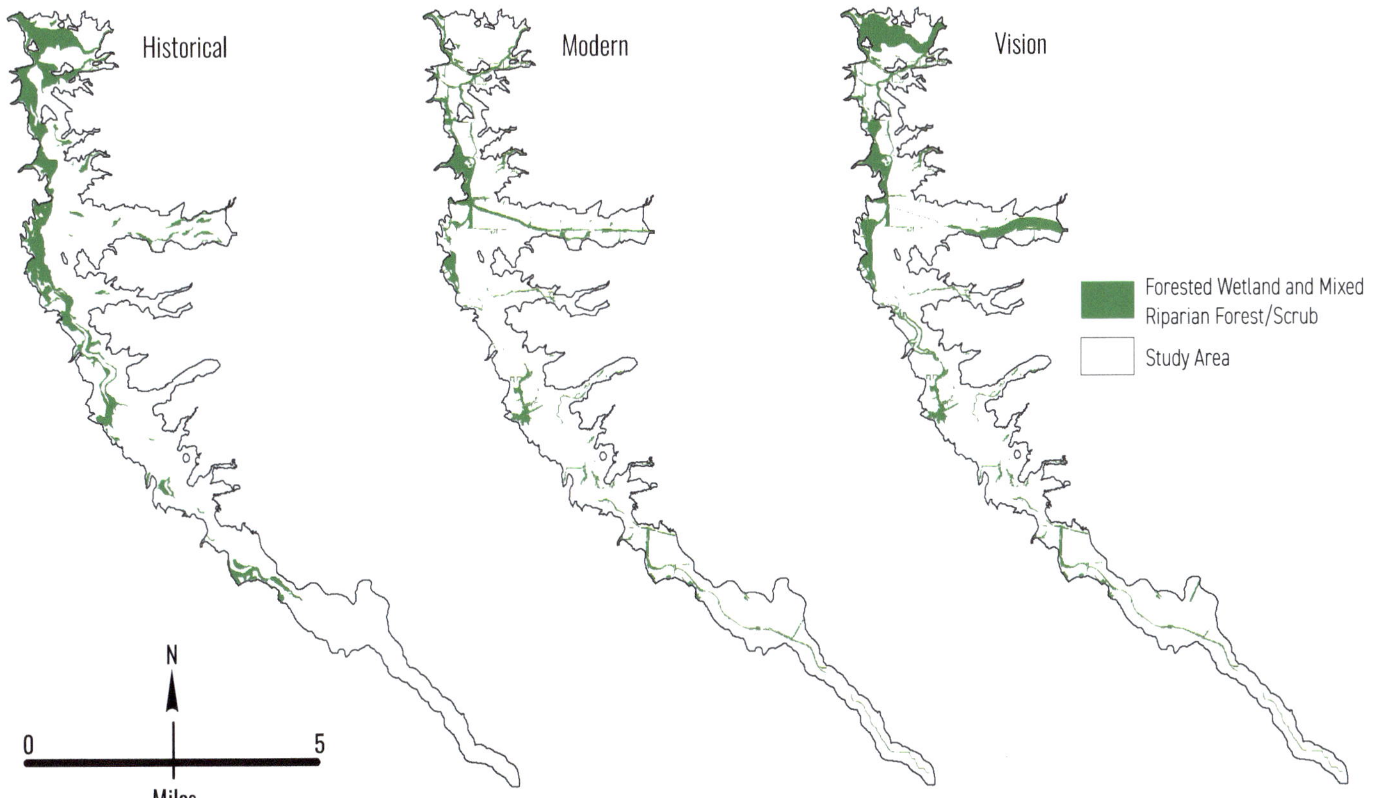

Historical

Modern

Vision

Forested Wetland and Mixed Riparian Forest/Scrub

Study Area

N

0 5

Miles

A

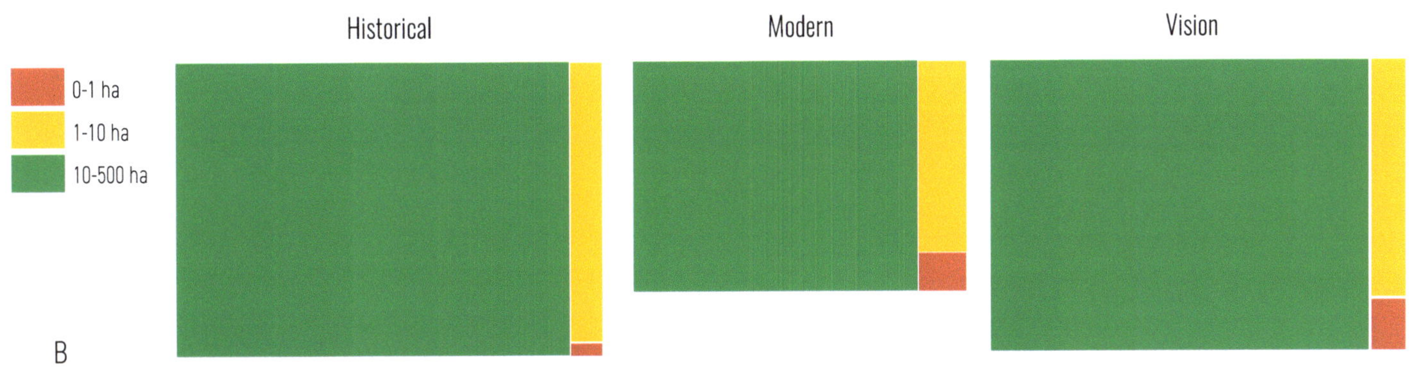

Distance to nearest large patch (m)	Area of riparian habitat (ha)		
	Historical	*Modern*	*Vision*
0-500	513	308	506
500-1000	11	17	9
1000-3000	13	14	14
Total	**538**	**339**	**528**

C

HABITAT CONFIGURATION

Greater extent of wetlands adjacent to channels.

Overall, the proportion of wetland areas adjacent to channels would increase with Vision implementation. The total channel length with adjacent valley freshwater marsh and wet meadow would have a similar degree of increase (67% and 65% respectively), with total channel length adjacent to wet meadow being close to the historical value. Due to conversion of drier land cover types to freshwater wetlands, the length of channels adjacent to oak savanna or woodland/ vernal pool complex/valley grassland areas would decrease by one-third and be similar to the historical value. Increasing adjacency of channels and wetlands would improve conditions for wildlife that occupy both habitat types and increase filtering of pollutants.

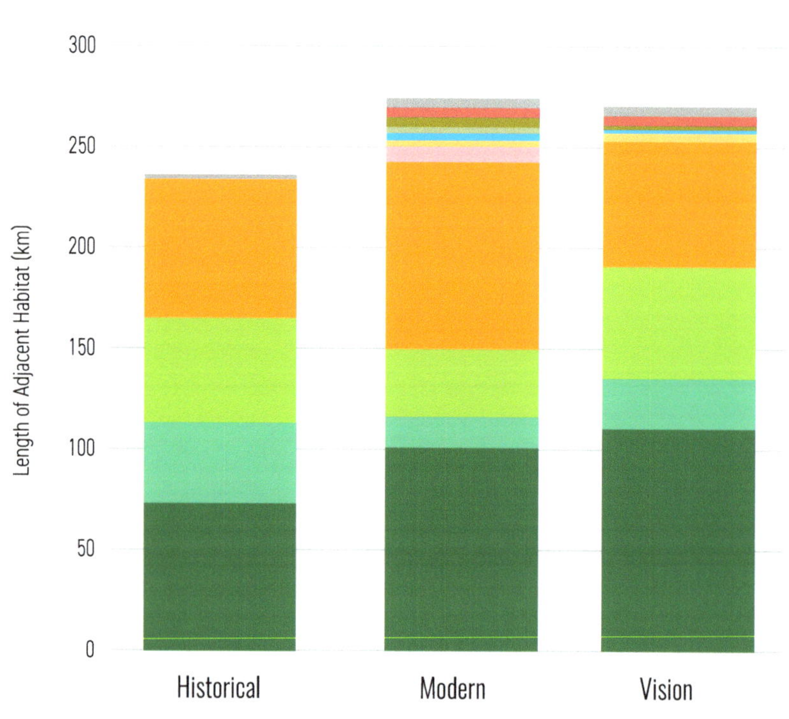

Length (km) of adjacent terrestrial habitats within a ~140 m buffer around contemporary channels.

HABITAT CONFIGURATION

Greater proportion of natural land cover types in terrestrial zones around wetlands.

The type and quality of landscape buffers around wetlands affect the quality of wetland habitats, and need to be considered in wetland design and restoration. Under the Vision, the proportion of wetlands surrounded by natural land cover types would increase from approximately 50% to 60%, mostly due to a five-fold increase in the area of riparian vegetation adjacent to wetlands. The area of both agricultural land and pasture/hayfield areas adjacent to wetlands would decrease modestly (15% and 4% respectively), but the area of developed/disturbed areas adjacent to wetlands would more than double, increasing from 170 to 380 ha (420 to 939 ac). The benefits of greater size and connectedness of wetland habitats overall, along with the greater proportion of wetlands surrounded by more natural habitats, would likely offset the increase in wetland adjacency to the developed/disturbed lands category.

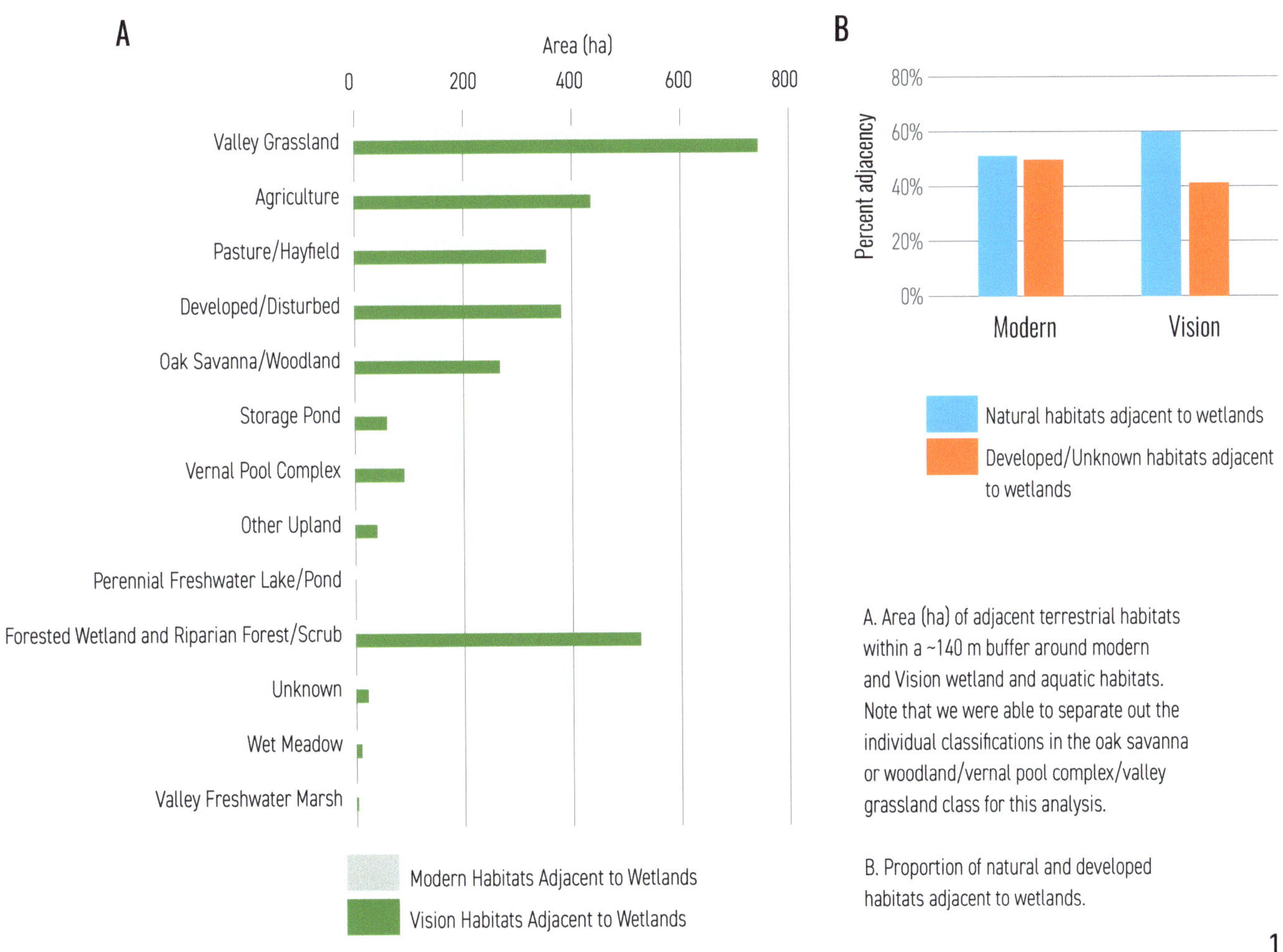

A. Area (ha) of adjacent terrestrial habitats within a ~140 m buffer around modern and Vision wetland and aquatic habitats. Note that we were able to separate out the individual classifications in the oak savanna or woodland/vernal pool complex/valley grassland class for this analysis.

B. Proportion of natural and developed habitats adjacent to wetlands.

Watershed-Scale Management for Supporting the Vision

The success of the restoration concepts presented in the Vision will depend in large part on active management within the Laguna and in its contributing watershed. Specifically, creating a resilient Laguna landscape that supports people and wildlife will require managing stormwater inflow, upstream groundwater recharge, sediment input and storage, nutrient input and storage, and invasive species. This section provides a high-level overview of the general types of management actions associated with these management themes that will need to be implemented both within and outside the Laguna to support the Vision concepts.

Modern Land Cover Types

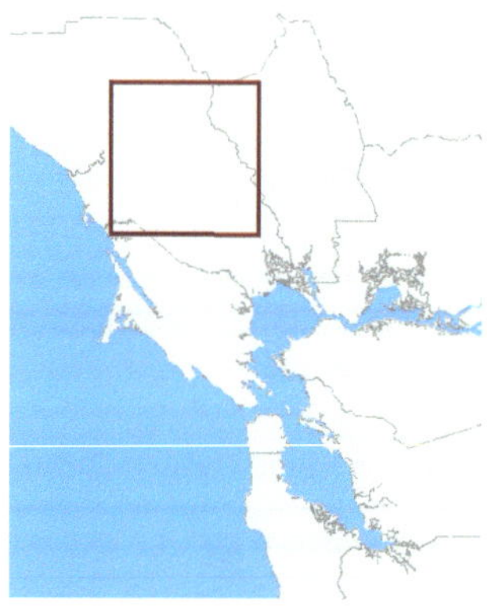

(Facing Page) Modern land cover types of the Laguna de Santa Rosa watershed. Source: Sonoma Vegetation Map 2015.

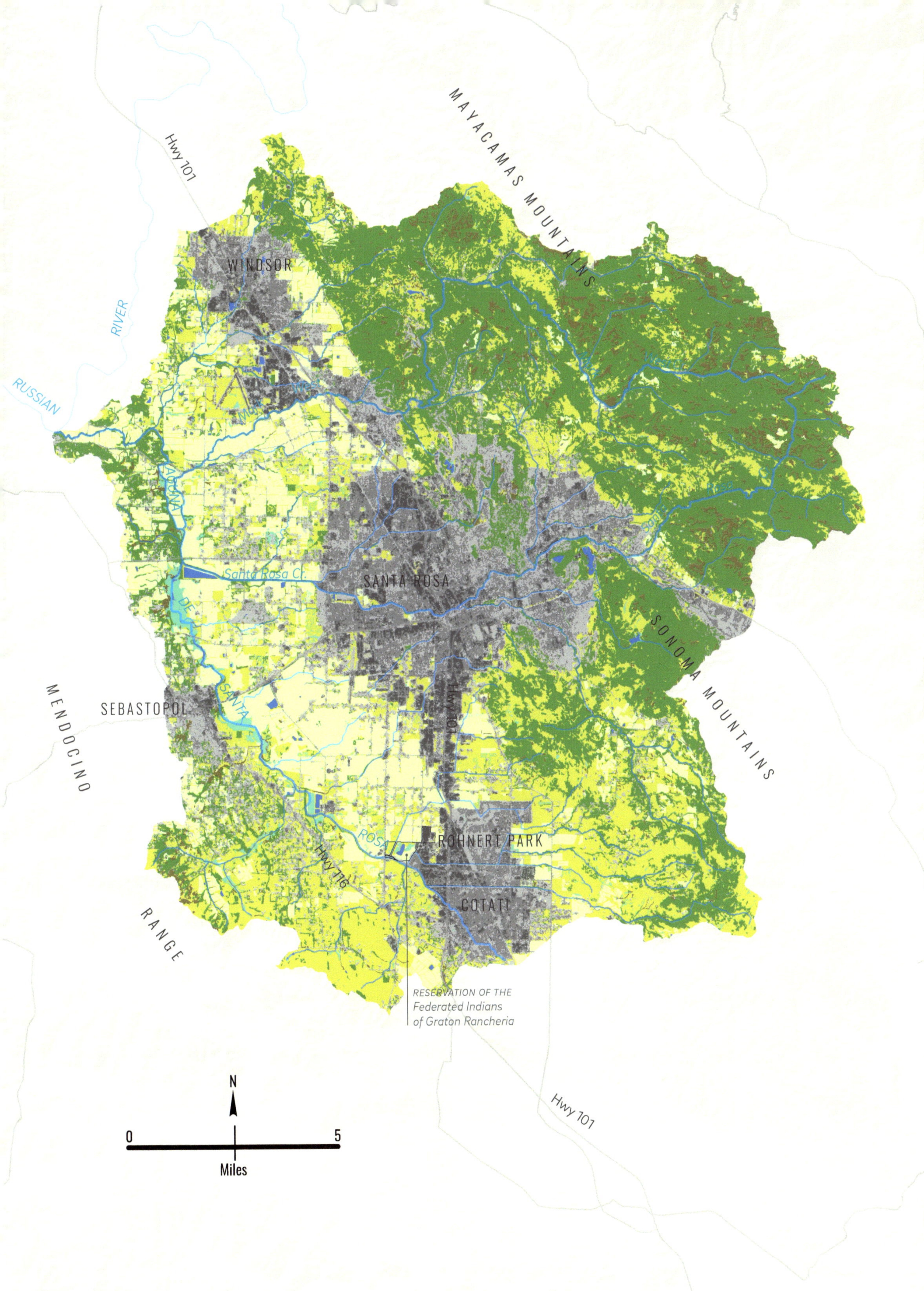
MAYACAMAS MOUNTAINS
Hwy 101
WINDSOR
RUSSIAN
RIVER
LAGUNA
Santa Rosa Cr.
SANTA ROSA
DE
SANTA
ROSA
SONOMA MOUNTAINS
MENDOCINO
RANGE
SEBASTOPOL
Hwy 101
ROHNERT PARK
Hwy 116
COTATI
RESERVATION OF THE
Federated Indians
of Graton Rancheria
Hwy 101
N
0
5
Miles

Flooding along the Laguna mainstem channel (March 13, 2010). Photo: Google Earth.

FLOW MANAGEMENT & GROUNDWATER RECHARGE

The Laguna's typical annual wintertime flooding extent is expanding due in large part to the rapid delivery of stormwater from the surrounding watershed (Curtis et al. 2013). Recent hydrologic modeling by the USGS indicates that the frequency of extreme flood flows to the Laguna from the surrounding watershed could be higher by the end of the century for a wetter or drier future (see Chapter 6). In addition, increased groundwater recharge is needed in the area to help meet potable water demand and to increase dry season flow into the Laguna for habitat support. Landscape-scale management of runoff will be needed to address stormwater flow to the Laguna, increase groundwater recharge upstream of the Laguna, and create a landscape that is resilient to a changing climate and provides benefits to both people and wildlife. Managing wintertime flows into the Laguna and increasing groundwater recharge can be addressed through several types of actions focused on source control, as well as capture and storage.

Source Control

Source control actions should focus on intercepting precipitation for small to moderate wintertime storm events (e.g., storm events with a recurrence interval of 5 years or less). On the hillslopes upstream of the alluvial fans and Santa Rosa Plain, the focus should be on re-establishing native evergreen woody vegetation (e.g., Douglas fir, coast live oak, bay laurel) where appropriate, to increase precipitation interception and decrease the amount of precipitation that falls directly on the ground surface, thereby decreasing runoff volumes. On the alluvial fans and the Santa Rosa Plain, the focus should be on both increasing native evergreen woody vegetation and implementing LID techniques to intercept flow. LID should include green infrastructure (GI) elements such as pervious pavement, rain gardens, tree-well planters, and bioswales. These elements should be installed on the areas of the Santa Rosa Plain identified as having high natural groundwater recharge potential (see Winzler & Kelly-GHD 2012). Tools like GreenPlanIT (greenplanit.sfei.org) can be used to site GI elements within these areas and determine the optimal combination of elements to achieve desired runoff reduction for the lowest cost.

Capture and Storage

Management actions should focus on capturing, storing, and infiltrating flood flows for all wintertime storm events. In the less developed areas on the alluvial fans and Santa Rosa Plain, the focus should be on increasing channel-floodplain connectivity through increasing channel elevation and/or decreasing floodplain elevation. The ideal sites have expansive undeveloped floodplains that could be inundated during small to moderate storm events and have high natural groundwater recharge potential. Within the Santa Rosa Creek watershed, SFEI (2017a) identifies and ranks a number of floodplain depressional wetlands in undeveloped areas that could capture flood flows and considerably decrease the amount of flood flow that reaches the Laguna. In the more developed areas on the alluvial fans and Santa Rosa Plain where open space is limited, the focus should be on multi-benefit floodwater detention facilities. These should include areas designed to store floodwaters during wintertime storm events, provide recreational opportunities when not flooded, and have habitat features that support a wide variety of native wildlife. Winzler & Kelly-GHD (2012) identifies a number of potential floodwater storage projects throughout the entire Laguna watershed that can help decrease flood flows to the Laguna and increase groundwater recharge.

Coarse sediment deposition at the transition from alluvial fan to Santa Rosa Plain, Copeland Creek at Snyder Lane bridge. Photo: SFEI.

SEDIMENT MANAGEMENT

The current sediment supply to the Laguna is thought to be much higher than it was before the onset of intensive development (PWA 2004b, Tetra Tech 2015b) and has the potential to increase in the future if climate change causes an increase in the frequency of large storm events. Establishing a Laguna landscape that supports people and wildlife into the future will require watershed-scale sediment management that includes erosion control and capture upstream of the Laguna as well as sediment removal within the Laguna. Management of sediment delivered to and deposited within the Laguna can be addressed through actions focused on source control, capture and storage, and active removal.

Source Control

Sediment source control actions should focus on reducing erosion throughout the Laguna watershed. In the hillslopes upstream of the alluvial fans and Santa Rosa Plain, the focus should be on reestablishing native woody vegetation on actively eroding slopes and steep headwater channel banks to decrease runoff velocity and shear stress and to help stabilize the ground surface. On the alluvial fans and the Santa Rosa Plain, the focus should be on controlling soil erosion from sparsely vegetated agricultural areas and stabilizing eroding channels that are producing a large amount of sediment. Soil erosion control actions should include increasing vegetation cover to intercept precipitation and reduce runoff velocity, thereby decreasing the amount of runoff and associated fine sediment transported to adjacent creeks and ultimately to the Laguna. Channel bank erosion control actions should include setting banks back to help decrease flow velocities and bank shear stress, and vegetating banks to help decrease flow velocity and shear stress and to stabilize banks. Actions to control channel incision should include large woody debris (LWD) installations that trap sediment and help build up bed elevations upstream over time.

Capture and Storage

The actions associated with capture and storage of wintertime flood flows described above have the additional benefit of capturing and depositing fine sediment outside of the Laguna. In less developed areas on the alluvial fans and Santa Rosa Plain, the focus should be on increasing channel-floodplain connectivity and getting sediment-laden flood flows onto floodplains where flows can slow and fine sediment can deposit. In the more developed areas on the alluvial fans and Santa Rosa Plain where open space is limited, the focus should be on multi-benefit floodwater detention facilities where fine sediment can deposit. The management approach for coarser sediment coming from the Sonoma and Mayacama Mountains and the Gold Ridge should include promoting deposition and storage on the historical alluvial fan deposits. As these deposits could cause decreased flood flow conveyance over time, they would most likely need to be located at the downstream end of Laguna tributary channels where increased flooding is a part of the restoration and management approach.

Removal

Even with watershed-wide erosion control and sediment capture and storage measures, sediment will need to be actively removed from the Laguna for flood conveyance, nutrient control, and overall habitat support. However, implementing the source control and capture and storage management actions described above for both flow and sediment will ideally lead to a decrease in the annual volume of sediment that will have to be removed from the mainstem Laguna channel and the lower reaches of Laguna tributaries.

NUTRIENT MANAGEMENT

Comprehensive management of nutrients through source reduction, interception, processing, and removal will be needed to reduce eutrophication at local and watershed scales. Important steps have already been taken to reduce phosphorus in the watershed through the recent adoption of a credit trading system (Kieser & Associates 2015, NCRWQCB 2018). However, co-management of nutrients rather than focusing on one 'limiting nutrient' will be key to success (Elser et al. 1990, Sloop et al. 2007, Conley et al. 2009, Harpole et al. 2011, Lewis et al. 2011, Jarvie et al. 2013, Dodds and Smith 2016, Sutula et al. 2018). Managing nutrients in the Laguna can be addressed through actions focused on source control, interception, and processing and removal.

Source Control

Reduction of point and non-point nutrient sources should be addressed in both urban and rural settings. Improvements in municipal wastewater treatment and septic systems can reduce urban and suburban sources. Meanwhile, implementation of urban and agricultural best management practices, including reduction in use of fertilizers, are major tools in nutrient source reduction (Bernhardt and Palmer 2007, Bernhardt et al. 2008).

Interception

Detention and retention of nutrients before they reach the Laguna can be achieved by increasing beneficial land cover types and landscape cover configurations that have high interception and cycling potential. Implementation of LID and Stormwater Control Measure practices in developed areas, such as bioretention gardens (i.e., rain gardens), infiltration wells and trenches, stormwater wetlands, permeable pavements, green roofs, vegetated buffer areas (strips or swales), sand filters, and water harvesting systems have been shown to greatly minimize runoff and pollutants to receiving streams (Dietz and Clausen 2008; Ahiablame et al. 2012; SFEI 2017a, 2017b). Agricultural best management practices that address nutrients typically involve intercepting nutrient-laden water and soil by dissipating flow energy, spreading flow, preventing erosion, and encouraging infiltration in farms and rangelands (Grismer et al. 2006, SCACO 2013).

Processing and Removal

Wetlands and riparian areas, including engineered wetlands adjacent to channels, can play an important role not only in intercepting, but also in processing nutrients, especially through denitrificaiton (the microbial process of moving nitrogen from soils and plant materials to the air) (Verhoeven et al. 2006, Mayer et al. 2007, Kaushal et al. 2008, Dosskey et al. 2010, Schipper et al. 2010, Peter et al. 2012, Baron et al. 2013, Baye, pers. comm, 2018). Additionally, nutrients can be stored in soils or plant biomass. Sequestration in soils can immobilize nutrients for long periods by burial in sediments deep enough to isolate them from biological mobilization, or by chemical transformation to biologically unavailable forms. In areas where nutrient-laden sediments and excess plant biomass (especially that of *L. hexapetala*) cause water quality problems, options for their physical removal should be considered, though the subsequent placement of the sediments and organic matter should be carefully planned and monitored to protect water quality.

Pastureland adjacent to the Laguna. Photo: SFEI.

INVASIVE VEGETATION MANAGEMENT

Invasive vegetation is a management issue throughout the Laguna watershed, and should be an important component of restoration efforts. There is no one recipe for invasive species control due to the variability of invasive species' biology, the sensitivity of desirable resources being managed, site accessibility, and interannual variability in site physical and ecological conditions (Grewell et al. 2016a). However, basic principles of weed management can be successful with a concerted effort. In the case of especially pernicious weeds common to the Laguna and its watershed like *L. hexapetala*, giant reed, perennial pepperweed, and Himalayan blackberry, a decadal approach to management is likely needed. Principles of weed management and specific control measures for *L. hexapetala* that could be successful are discussed below.

Ludwigia hexapetala growth in the Laguna. Photo: Julian Meisler, Laguna de Santa Rosa Foundation.

Principles of Weed Management

Several best management practices or principles of weed management can be applied in the Laguna and its watershed. The following list reiterates principles of weed management laid out in Honton and Sears (2006), Grewell et al. (2016a), and DiTomaso et al. (2017) (Table 7-1) .

Table 7-1. Principles of Weed Management

Category	Description
Prevention	Prevention of sales of invasive plants, and public education reduces intentional and unintentional introductions of plant material.
Early Detection	Monitoring for new invaders and re-establishing populations of invasive plants allows for treatment of smaller, more manageable patches.
Seasonal Removal	Removal treatments should ideally be before the target plant has flowered or set seed, thereby reducing that year's contribution to local seed banks. Test and implement integrated weed management methods to suppress and remove biomass. Manual removal is feasible for small infestations, but mechanical removal is usually necessary for well-established populations. When using mechanical removal, care should be taken to capture root and shoot fragments so they do not spread and establish elsewhere.
Seasonal Herbicide Use	Herbicides can be used as a method of last resort, where drinking water uses, sensitive species, and legal or permit requirements make them an option. Herbicides have been effective in reducing invasive weed populations in both terrestrial and aquatic settings. As with mechanical removal, timing of treatments should occur before flower or seed set.
Integrated Weed Management	Use of a combination of biological, chemical, hydrological, and other methods, based on the biology and ecology of the weed species can have the best outcome, while reducing reliance on pesticides.
Post-Treatment Annual Maintenance Management	As part of a long-term effort, managers should return to treated sites to look for resprouting from unremoved rhizomes, and emergence of new plants from persistent seed banks. Additionally, reintroduction of native species in treated areas is recommended.
Biological Control	For very widespread and pernicious weeds, biological control through the introduction of pathogens or competitors, or the use of grazing within an integrated weed management strategy, may be desirable. Successful biological control requires careful research and experimentation before widespread application.

Control Strategies for Ludwigia hexapetala

As one of the most problematic and widespread weeds in the Laguna, control strategies specifically addressing *L. hexapetala* merit special attention. Earlier efforts to remove *L. hexapetala* in the Laguna have provided valuable lessons, and recent research has provided insights into its control. The following is a set of *L. hexapetala*-specific management actions.

Control of nutrient concentrations and loads watershed-wide. Biomass of *L. hexapetala* increases with elevated soil nutrient availability, because it draws most of its nutrients from the sediments, not the water column (Grewell et al. 2016b). Dredging and removal of nutrient-rich soils, and reducing nutrient concentrations and loads, should help reduce enriched sediment conditions conducive to *L. hexapetala* growth.

Regional management. Re-introductions of *Ludwigia* spp. from seed and root fragments by water connection between the Laguna and its tributaries, as well as the Russian River, are possible (Okada et al. 2009, Skaer Thomason et al. 2018a, Grewell 2019). This highlights the need for coordinated regional weed management.

Remove biomass based on life stage of the weed. Biomass removal should be timed based on points in the life cycle of the species when control is most effective. In the case of *L. hexapetala*, control efforts should take place spring-summer into the flowering life stage, but before the plants have produced seed capsules and have exponentially increased biomass. Late summer or fall treatments will not be effective. *L. hexapetala* seeds are already present in the Laguna soils, so treatment of new growth will likely be necessary (Grewell and Futrell 2009, Grewell et al. 2016a, 2019).

Monitor water quality when using herbicides. Remove excessive biomass before herbicide application to avoid severe oxygen depletion in the water column due to excess decomposing material, and plan for potentially costly monitoring of water quality as part of the treatments (Meisler 2008, 2009).

Plant native competitor species. In gaps left by removal of weeds, native competitor species should be planted. Solely reducing weedy biomass without also planting desirable species can invite re-emergence of the targeted weed, or leave space for new invaders.

Commit to annual maintenance management. Initial treatments can be effective for a few years, but re-emergence from root fragments and the seed bank, or by re-introduction, is likely. A commitment of years to decades will likely be needed to manage *L. hexapetala*.

Adaptively manage for native species within invaded areas. Patches of native wetland species have managed to persist in certain areas otherwise highly invaded by *L. hexapetala* within the Laguna. One hypothesis for this is that the native stands of vegetation are found at slightly higher elevations than areas invaded by *L. hexapetala*. Another explanation could be that the persistent native plants are genetically adapted to surviving under the pressures of invasion. *L. hexapetala* in the Laguna has also been observed to be less prevalent under shady canopies and in deeper waters (Baye 2008). Establishing native competitors that shade channels and lakes, such as dense stands of the tall emergent plants like river bullrush (*Bolboschoenus fluvialitis*), tules (*Schoenoplectus acutus*), and cattail (*Typha* spp.), as well as riparian trees such as willow (*Salix* spp.) and box-elder (*Acer negundo*) could help control *L. hexapetala* (Skaer Thomason et al. 2018, Baye 2008). Additionally, hydrological isolation could be a factor in controlling areas that remain uninvaded by *L. hexapetala*. Experimental investigations of these factors, paired with hypothesis-driven monitoring, could examine these patterns and guide future wetland restoration efforts that incorporate management of *L. hexapetala*. §

Ludwigia hexapetala blooming in the Laguna. Photo: Brenda Grewell.

8 Key Considerations and Known Challenges

This Vision identifies restoration opportunities for the Laguna that can help promote biodiversity and enhance ecosystem services for wildlife and people. Implementation of any given project developed from these opportunities will need to consider a range of factors and known challenges associated with coordinating existing land uses and other landscape management efforts, funding for restoration planning and construction, and permitting of restoration efforts. Below is a discussion of some of the key considerations and known challenges associated with moving Laguna restoration efforts forward.

Kayaking in the Laguna de Santa Rosa. Photo: SFEI.

LANDOWNER COORDINATION and OUTREACH • Much of the land in and around the Laguna is privately held, so implementing projects associated with many of the opportunities shown in the Vision will require private landowner willingness. Landowner and public outreach will be needed to accomplish any of these restoration opportunities. During the public outreach meetings for this project, many local landowners expressed interest in helping restore the Laguna. The project team will look to recent restoration efforts in the region (e.g., Dry Creek downstream of Lake Sonoma) to determine the best approach for continued coordination with local landowners as restoration project ideas for the Laguna continue to develop. Ongoing consultation with the Federated Indians of Graton Rancheria will also be a critical step to ensure that restoration projects near the Tribe's reservation and other culturally significant lands do not impact the Tribe's interests.

FUNDING FOR RESTORATION PROJECTS • A variety of funding sources is available for completing restoration projects within the Laguna, including federal, state, and local government restoration grants, and water quality credit trading administered by the NCRWQCB. Securing funding that covers costs from design through implementation and monitoring will be essential for restoration success.

EXISTING INFRASTRUCTURE • Implementing restoration actions developed from this Vision will require modification, relocation, or removal of existing infrastructure. For example, the Vision calls for relocation of flood control levees, redesign of bridges, and the relocation of wastewater treatment facilities. The anticipated benefits associated with these actions will need to be assessed through detailed technical analyses during project feasibility and design phases.

SEQUENCING OF RESTORATION PROJECTS • The sequencing of restoration actions will be important to attain desired long-term ecosystem benefits. For example, the success of many restoration actions is predicated on changes to the supply of flow, sediment, and/or nutrients. The proper sequencing of projects to meet desired restoration goals will need to be determined, which will ultimately help prioritize project funding requests.

COORDINATION WITH OTHER MANAGEMENT EFFORTS • There are currently several state and local agencies developing new or updated approaches for managing the Laguna and the surrounding landscape to better support wildlife and people. These include the NCRWQCB developing a nutrient credit trading program in the Laguna watershed to help meet the established phosphorus TMDL (Kieser & Assoc. 2015, NCRWQCB 2018); Permit Sonoma working to update the Russian River Municipal Separate Storm Sewer System permit (MS4) that details best management practices to improve water quality in receiving water bodies; and the Santa Rosa Plain Groundwater Sustainability Agency developing a Groundwater Sustainability Plan that will establish standards to ensure the sustainable use of groundwater within the Santa Rosa groundwater basin. The restoration projects inspired by this Vision will need to be developed in coordination with these, and other regional efforts to ensure that restoration projects generally support these regional management goals and, where possible, provide a clear way to help meet the management goals of these regional efforts. §

Morning mist in the Laguna de Santa Rosa. Photo: SFEI.

9 Synthesis and Next Steps

The Laguna de Santa Rosa is a vibrant landscape that has experienced a range of changes over the past two centuries, impacting both people and wildlife. Moving forward, the Laguna landscape will continue to change as the local population increases and shifts in climate bring higher air temperatures and more extreme storms. Sustaining and enhancing the many ecosystem services the landscape provides will require coordinated management and restoration actions both within the Laguna and in its contributing watershed. The Vision presented here is intended to be a road map for management and restoration actions that improve ecosystem resilience over the short- and long-term.

Synthesis

This Vision seeks to support ecosystem management in the Laguna by contributing to an understanding of how the landscape in the Laguna has functioned over time. This understanding is informed by the dimensions of landscape resilience identified in the Landscape Resilience Framework (Beller et al. 2015), starting with analyses of the unique setting of the Laguna, and how this setting influences key physical processes. The Vision concepts then address the landscape configurations and processes that in turn influence the ecosystem services and wildlife the Laguna supports. As an integral part of the Laguna landscape, people influence its processes and land uses through time and at all scales. Through restoration and management, we can support and enhance ecosystem functions in the Laguna that benefit both wildlife and people. The following summarizes how this Vision addresses dimensions of landscape resilience.

SETTING AND PROCESS

The unique physical setting of the Laguna contributes to the character of the habitats and wildlife it supports. The slow transition from lower to higher elevation areas surrounding the Laguna support a variety of wetlands and uplands. Year-round access to groundwater, along with the unique mix of soil properties, support a diverse range of plant communities. All these shape the incredible diversity of the Laguna's wetlands and uplands, and form the constraints and opportunities for building landscape resilience in the Laguna. People have influenced this setting over time, with major changes to the landscape. Novel land cover types have been created in our cities, agriculture, and associated infrastructure, and new plant and animal species have been introduced. The Vision concepts that increase wetland and riparian habitat extent, and that reconfigure land cover, in part address the effects of these landscape changes by increasing wetland and riparian habitat extent, complexity and connectedness.

Patterns in the key processes at play in the Laguna watershed, including hydrologic, sediment, and nutrient dynamics, have changed over time. Water and sediment delivery to the Laguna, as well as concentrations and loads of nutrients, have increased. The undesirable effects of these changes can be reduced through implementation of Vision concepts including increases in wetland land cover types; thoughtful configuration of these wetlands on the landscape to support habitats as well as improve water quality; and informed management of water, sediment, nutrients, and invasive species across the landscape. Together these measures can support ecosystem services such as pollution filtration and aesthetic values for people, and improve habitat for the Laguna's native plants and wildlife.

CONNECTIVITY, DIVERSITY, REDUNDANCY, AND SCALE

Alongside major changes to key processes that influence the setting for wildlife and people, land cover and land use have also changed dramatically, most notably in the loss and fragmentation of wetland habitats. The Vision highlights opportunities for restoring well-connected, diverse, and complex areas of open water, freshwater marsh, wet meadow, and woody riparian areas that could provide a wide range of ecosystem services over time. The sustainable use of working landscapes and green design of urban landscapes that promote sediment and nutrient capture and enhance habitats will further support ecological resilience in the Laguna.

As restoration progresses in the Laguna, ensuring that habitats are of high quality, and are redundant at a variety of spatial scales, will allow resident and migratory wildlife populations to adapt and acclimate to future climate stresses. Vision concepts that lead to an increase in the connectedness of high-quality habitats will support biodiversity both within the Laguna and at watershed and regional scales. Vision concepts that lead to decreased wetland and riparian fragmentation will benefit plants and animals within their local home ranges. Resident birds, reptiles, and amphibians, as well as mammals such as bats, bobcats and skunks that typically have relatively small home ranges for daily foraging, resting, and reproduction would greatly benefit from decreased habitat fragmentation. Wildlife with wide ranges that go outside the Laguna such as salmonids, migratory birds, beaver, river otter, deer, elk, and mountain lion will also benefit from improvements in Laguna habitat quality during portions of their lives for movement, dispersal, breeding and feeding.

PEOPLE

The role of people—residents, landowners, land and water managers—will be key to achieving positive outcomes in the Laguna. The Laguna will continue to be a place where people live and work, in both urban and rural settings. Over the last few decades, many people have already contributed to improving the Laguna landscape by sustaining native plants and animals, preserving cultural ties to the land, and educating new generations of young people. We now have an opportunity to renew our commitment to a shared vision for enhancing and caring for the Laguna through collaborating to restore habitats that together achieve a well-configured, wildlife-friendly, resilient landscape where people can thrive.

Education program in the Laguna. Photo: Laguna Foundation.

Next Steps

This Vision is the first step in the larger Laguna-Mark West Creek Watershed Master Restoration Planning Project. With the Vision completed, the project focus turns to developing a Restoration Plan for the Laguna and designs for priority projects identified in the Restoration Plan.

RESTORATION PLAN

The Laguna Restoration Plan will build from the Vision, describing near-term restoration targets and project concepts that help meet the targets. The overall goal will be to increase the extent of wetland habitat within the next few decades. The specific targets will be determined in close coordination with the project technical advisors, stakeholders, and local landowners using the difference between the modern habitat extent and the habitat extent associated with the Vision (i.e., the long-term restoration goal) as guidance. The suite of project concepts shown in the Restoration Plan will be developed directly from the restoration opportunities shown in the Vision. The Restoration Plan will provide details about the process for developing the project concepts, the habitat features within each project concept, each project concept's contribution toward meeting the near-term restoration targets, and the recommended order of implementation. Additionally, the Restoration Plan will identify key data and knowledge gaps, and research opportunities that will help achieve the restoration concepts.

RESTORATION PROJECT DESIGNS & PERMITTING

One or two of the restoration project concepts in the Restoration Plan identified as high priority will have design plans developed and corresponding permit applications submitted to the appropriate regulatory agencies. For each high priority project concept selected, design plan development will include: an assessment of site physical functioning; hydraulic modeling of existing conditions; an assessment of habitat enhancement alternatives based on model results and established restoration targets; conceptual (30%) designs for the preferred alternative identified through conversations with the TAC, stakeholders, and CDFW; intermediate (65%) designs for the preferred alternative that incorporates feedback from the TAC, stakeholders, and CDFW; and a basis-of-design report that summarizes site conditions, restoration design elements, and monitoring recommendations. Permit application materials will include wildlife and plant resource survey reports, jurisdictional wetland delineation reports (as appropriate), and CEQA documents. Sonoma Water will submit these materials as part of permit applications to the US Army Corps of Engineers, National Marine Fisheries Service, USFWS, and CDFW. §

10 References

Aalto, R. 2004. "Report on Sediment Accumulation Rates for a Floodplain of the Laguna de Santa Rosa, CA, Determind with 210Pb Geochronology." U.S. Army Corps of Engineers.

Ackerly, D.D., Cornwell W.K., Weiss, S.B., Flint L.E., and Flint. A.L. 2015. "A Geographic Mosaic of Climate Change Impacts on Terrestrial Vegetation: Which Areas Are Most at Risk?" *PLOS ONE* 10 (6): e0130629. https://doi.org/10.1371/journal.pone.0130629.

Ahiablame, L.M., Engel B.A., and Chaubey, I. 2012. "Effectiveness of Low Impact Development Practices: Literature Review and Suggestions for Future Research." *Water, Air, & Soil Pollution* 223 (7): 4253–73. https://doi.org/10.1007/s11270-012-1189-2.

Anderson, M.K. 2005. *Tending the Wild: Native American Knowledge and the Management of California's Natural Resources*. Univ of California Press.

Asner, G.P., Brodrick, P.G., Anderson, C.B., Vaughn, N., Knapp, D.E., and Martin, R.E. 2016. "Progressive Forest Canopy Water Loss during the 2012–2015 California Drought." *Proceedings of the National Academy of Sciences* 113 (2): E249–55. https://doi.org/10.1073/pnas.1523397113.

Baker, M.S. 1899. "Consortium of California Herbaria, Record for Potamogeton Ilinoensis from 'Near Sebastopol.'" University Herbarium, UC Berkeley.

———. 1900. "Consortium of California Herbaria, Record for Potamogeton Ilinoensis." University Herbarium, UC Berkeley.

Barbour, M.G, Solomeshch, A.I., Buck, J.J., Holland, R.F., Witham, C.W., MacDonald, R.L., Starr, S.L., and Lazar, K.A. 2007. "Classification, Ecological Characterization, and Presence of Listed Plant Taxa of Vernal Pool Associations in California." USFWS Agreement/Study No. 814205G238. University of California, Davis.

Baron, J.S., Hall, E.K., Nolan, B.T., Finlay, J.C., Bernhardt, E.S., Harrison, J.A., Chan, F., and Boyer, E.W. 2013. "The Interactive Effects of Excess Reactive Nitrogen and Climate Change on Aquatic Ecosystems and Water Resources of the United States." *Biogeochemistry* 114 (1): 71–92. https://doi.org/10.1007/s10533-012-9788-y.

Barrett, S.A. 1908. *The Ethno-Geography of the Pomo and Neighboring Indians*. University of California Publications in American Archaeology and Ethnology. Berkeley, CA.

Baumgarten, S.A., Grossinger, R.M., and Beller, E.E. 2017. "Historical Ecology and Landscape Change in the Central Laguna de Santa Rosa." SFEI Publication #820, San Francisco Estuary Institute, Richmond, CA.

Baumgarten, S., Beller, E., Grossinger, R., Striplen, C., Brown, H., Dusterhoff, S., Salomon, M., and Askevold, R. 2014. "Historical Changes in Channel Alignment along Lower Laguna de Santa Rosa and Mark West Creek." SFEI Publication #715, San Francisco Estuary Institute, Richmond, CA.

Baye, P.B. 2008. "Summary of Field Trip Discussion Topics: Laguna de Santa Rosa Floodplain Wetlands, July 15, 2008," July 24, 2008.

———. 2018. "Laguna de Santa Rosa: basis for co-management of N and P to address eutrophication impacts with co-benefits for aquatic and wetland ecosystems." December 28, 2018.

Beach, R.F. 2002. *History of the Development of the Water Resources of the Russian River*. Sonoma County Water Agency.

Beller, E.E., Spotswood, E.N., Robinson, A.H., Anderson, M.G., Higgs, E.S., Hobbs, R.J., Suding, K.N., Zavaleta, E.S., Grenier, L.J., and Grossinger, R.M. 2019. "Building Ecological Resilience in Highly Modified Landscapes." *BioScience* 69 (1): 80–92. https://doi.org/10.1093/biosci/biy117.

Beller, E., Robinson, A., Grossinger, R.M., and Grenier, L. 2015. "Landscape Resilience Framework: Operationalizing Ecological Resilience at the Landscape Scale." SFEI Publication #752, San Francisco Estuary Institute, Richmond, CA.

de Bello, F., Lavorel, S., Díaz, S., Harrington, R., Cornelissen, J.H.C, Bardgett, R.D., Berg, M.P., et al. 2010. "Towards an Assessment of Multiple Ecosystem Processes and Services via Functional Traits." *Biodiversity and Conservation* 19 (10): 2873–93. https://doi.org/10.1007/s10531-010-9850-9.

Bernhardt, E.S. and Palmer, M.A., 2007. Restoring streams in an urbanizing world. Freshwater biology, 52(4), pp.738-751.

Bernhardt, E.S., Band, L.E., Walsh, C.J. and Berke, P.E., 2008. Understanding, managing, and minimizing urban impacts on surface water nitrogen loading. Annals of the New York Academy of Sciences, 1134(1), pp.61-96.

Bernhardt, E.S., Palmer, M.A. 2011. "River Restoration: The Fuzzy Logic of Repairing Reaches to Reverse Catchment Scale Degradation." *Ecological Applications* 21 (6): 1926–31. https://doi.org/10.1890/10-1574.1.

Best, C., Howell, J.T., Knight, W., Knight, I., and Wells, M. 1996. "A Flora of Sonoma County." Sacramento, CA: California Native Plant Society.

Bowers, AB. 1866. "Map of Sonoma County, California." *Courtesy of David Rumsey Map Collection*.

Bowman, D.M.J.S., Balch, J., Artaxo, P., Bond, W.J., Cochrane, M.A., D'Antonio, C.M., DeFries, R., et al. 2011. "The Human Dimension of Fire Regimes on Earth." *Journal of Biogeography* 38 (12): 2223–36. https://doi.org/10.1111/j.1365-2699.2011.02595.x.

Brown, M., and Dinsmore, J.J. 1986. "Implications of Marsh Size and Isolation for Marsh Bird Management." *The Journal of Wildlife Management* 50 (3): 392. https://doi.org/10.2307/3801093.

Bulger, J.B., Scott, N.J., and Seymour, R.B. 2003. "Terrestrial Activity and Conservation of Adult California Red-Legged Frogs Rana Aurora Draytonii in Coastal Forests and Grasslands." *Biological Conservation* 110 (1): 85–95. https://doi.org/10.1016/S0006-3207(02)00179-9.

Butchart, S.H.M., Walpole, M., Collen, B., van Strien, A., Scharlemann, J.P.W., Almond, R.E.A., Baillie, J.E.M., et al. 2010. "Global Biodiversity: Indicators of Recent Declines." *Science* 328 (5982): 1164–68. https://doi.org/10.1126/science.1187512.

Butkus, S. 2010. "Nutrient Loading Estimates for Laguna TMDL Source Analysis." Santa Rosa, CA: North Coast Regional Water Quality Control Board.

———. 2011a. "Constructing Stream Flow Rating Power Equations for the Pre-Settlement Lakes in the Laguna de Santa Rosa Watershed." Santa Rosa, CA: North Coast Regional Water Quality Control Board.

———. 2011b. "Development of the Laguna de Santa Rosa Watershed Pre-European Settlement Spatial Data Model." Santa Rosa, CA: North Coast Regional Water Quality Control Board.

Cardwell, G.T. 1958. "Geology and Ground Water in the Santa Rosa and Petaluma Valley Areas, Sonoma County, California." USGS Numbered Series 1427. Water Supply Paper. U.S. Govt. Print. Off.,. http://pubs.er.usgs.gov/publication/wsp1427.

CBD (Convention on Biological Diversity). 2014. "Pathways of Introduction of Invasive Species, Their Prioritization, and Management." Montreal, Canada: United Nations Environmental Programme, Convention on Biological Diversity.

CDFG (California Department of Fish and Game). 2004. "Recovery Strategy for California Coho Salmon." Species Recovery Strategy 2004–1. Sacramento, CA: California Department of Fish and Game.

CDFW (California Department of Fish and Wildlife). n.d. "Russian River Tule Perch (Hysterocarpus Traskii Pomo)." California Department of Fish and Game.

———. 2015. "Sonoma County Water Agency Climate Vulnerability Assessment and Adaptation Work Plan." Work Plan.

Chappelle, C., McCann, H., Jassby, D., Schwabe, K., Szeptycki, L. 2019. "Managing Wastewater in a Changing Climate." San Francisco, CA: Public Policy Institute of California.

Chornesky, E.A., Ackerly, D.D., Beier, P., Davis, F.W., Flint, L.E., Lawler, J.J., Moyle, P.B., et al. 2015. "Adapting California's Ecosystems to a Changing Climate." *BioScience* 65 (3): 247–62. https://doi.org/10.1093/biosci/biu233.

City of Santa Rosa, County of Sonoma, and Sonoma County Water Agency. 2013. "Santa Rosa Citywide Creek Master Plan." Santa Rosa, CA: Santa Rosa City Council.

Cloern, J.E. 2001. "Our Evolving Conceptual Model of the Coastal Eutrophication Problem." *Marine Ecology Progress Series* 210 (January): 223–53. https://doi.org/10.3354/meps210223.

Cluer, B., Thorne, C. 2014. "A Stream Evolution Model Integrating Habitat and Ecosystem Benefits." *River Research and Applications* 30 (2): 135–54. https://doi.org/10.1002/rra.2631.

CNRA (California Natural Resources Agency). 2018. Safeguarding California Plan: California's Climate Adaptation Strategy. January 2018. https://resources.ca.gov/CNRALegacyFiles/docs/climate/safeguarding/update2018/safeguarding-california-plan-2018-update.pdf

Community Foundation Sonoma County. 2010. Biodiversity Action Plan: Priority Actions to Preserve Biodiversity in Sonoma County. Prepared for Community Foundation Sonoma County, Sonoma County Water Agency.

Conley, D.J., Paerl H.W., Howarth R.W., Boesch D.F., Seitzinger S.P., Havens K.E., Lancelot C., and Likens G.E. 2009. "Controlling Eutrophication: Nitrogen and Phosphorus." *Science* 323 (5917): 1014–15. https://doi.org/10.1126/science.1167755.

Coombs, J.S., and Melack, M.M. 2013. "Initial Impacts of a Wildfire on Hydrology and Suspended Sediment and Nutrient Export in California Chaparral Watersheds." *Hydrological Processes* 27 (26): 3842–51. https://doi.org/10.1002/hyp.9508.

Cummings, J. 2003a. "The Awful Offal of Sebastopol." Petaluma, CA. Sonoma State University Library, North Bay Digital Collections.

———. 2003b. "Crystal Lauging Waters - Historical Glimpses of the Laguna de Santa Rosa." Petaluma, CA. Sonoma State University Library, North Bay Digital Collections.

———. 2004. "Draining and Filling the Laguna de Santa Rosa." Petaluma, CA. Sonoma State University Library, North Bay Digital Collections.

———. 2006. "Early Sebastopol: Part IV - 'Sprightly Sebastopol' - a 'Lively Burg'". Petaluma, CA. Sonoma State University Library North Bay Digital Collections.

Curtis, J.A., Flint L.E., and Hupp, C.R. 2013. "Estimating Floodplain Sedimentation in the Laguna de Santa Rosa, Sonoma County, CA." *Wetlands* 33 (1): 29–45. https://doi.org/10.1007/s13157-012-0350-4.

Dahl, K., Licker, R., Abatzoglou, J.T., and Declet-Barreto, J. 2019. "Increased Frequency of and Population Exposure to Extreme Heat Index Days in the United States during the 21st Century." *Environmental Research Communications* 1 (7): 075002. https://doi.org/10.1088/2515-7620/ab27cf.

Daily Alta California. 1866. "The Storm in the Interior." December 22, 1866. *Courtesy of California Digital Newspaper Collection.*

———. 1888. "A City's Sewage." October 23, 1888. *Courtesy of California Digital Newspaper Collection.*

Das, A.J., Stephenson, N.L., Flint, A., Das, T., and van Mantgem, P.J. 2013. "Climatic Correlates of Tree Mortality in Water- and Energy-Limited Forests." *PLOS ONE* 8 (7): e69917. https://doi.org/10.1371/journal.pone.0069917.

Davis, P.R. 1887. "Plan of Land (Formerly) Belonging to the Nicolas Carriger Estate, Sonoma Co. California Being a Portion of the San Miguel Rancho." Curtis and Associates, Inc.

Dawson, A., and Sloop, C. 2010. "Laguna de Santa Rosa Historical Hydrology Project Headwaters Pilot Study." Eldridge, CA: Sonoma Ecology Center.

Denner, S. 2002. Transcript of oral interview of Stan Denner recorded by Jane Nielson on August 30, 2002.

Dettinger, M.D. 2013. "Atmospheric Rivers as Drought Busters on the U.S. West Coast." *Journal of Hydrometeorology* 14 (6): 1721–32. https://doi.org/10.1175/JHM-D-13-02.1.

Díaz-Delgado, R., Lloret, F., Pons, X. and Terradas, J., 2002. Satellite evidence of decreasing resilience in Mediterranean plant communities after recurrent wildfires. Ecology, 83(8), pp.2293-2303.

Dietz, M.E. and Clausen, J.C., 2008. Stormwater runoff and export changes with development in a traditional and low impact subdivision. Journal of environmental management, 87(4), p.560.

Diffenbaugh, N.S., Swain, D.L., and Touma, D. 2015. "Anthropogenic Warming Has Increased Drought Risk in California." *Proceedings of the National Academy of Sciences* 112 (13): 3931–36. https://doi.org/10.1073/pnas.1422385112.

DiTomaso, J.D., Bell, C.E., and Wilen, C.A. 2017. "UC IPM Pest Notes: Invasive Plants." 74139. Oakland, CA: University of California Integrated Pest Management Program.

Dodds, W.K, Smith V.H., and Lohman. K. 2002. "Nitrogen and Phosphorus Relationships to Benthic Algal Biomass in Temperate Streams." *Canadian Journal of Fisheries and Aquatic Sciences* 59 (5): 865–74. https://doi.org/10.1139/f02-063.

Dodds, W., and Smith, V. 2016. "Nitrogen, Phosphorus, and Eutrophication in Streams." *Inland Waters* 6 (2): 155–64. https://doi.org/10.5268/IW-6.2.909.

Dosskey, M.G., Vidon, P., Gurwick, N.P., Allan, C.J., Duval, T.P., and Lowrance, R. 2010. "The Role of Riparian Vegetation in Protecting and Improving Chemical Water Quality in Streams1." *JAWRA Journal of the American Water Resources Association* 46 (2): 261–77. https://doi.org/10.1111/j.1752-1688.2010.00419.x.

Dyer, E.H. 1861. "Plat of the Llano de Santa Rosa, Finally Confirmed to Joaquin Carillo. Surveyed under Instructions from the U.S. Surveyor General." Bureau of Land Management.

Ehrenfeld, J.G. 2010. "Ecosystem Consequences of Biological Invasions." *Annual Review of Ecology, Evolution, and Systematics* 41 (1): 59–80. https://doi.org/10.1146/annurev-ecolsys-102209-144650.

Elser, J.J., Marzolf E.R., and Goldman, C.R. 1990. "Phosphorus and Nitrogen Limitation of Phytoplankton Growth in the Freshwaters of North America: A Review and Critique of Experimental Enrichments." *Canadian Journal of Fisheries and Aquatic Sciences* 47 (7): 1468–77. https://doi.org/10.1139/f90-165.

Essl, F., Bacher, S., Blackburn, T.M., Booy, O., Brundu, G., Brunel, S., Cardoso, A., et al. 2015. "Crossing Frontiers in Tackling Pathways of Biological Invasions." *BioScience* 65 (8): 769–82. https://doi.org/10.1093/biosci/biv082.

Ferriter, M.M., Quantifying Post-Wildfire Vegetation Regrowth in California since Landsat 5. https://nature.berkeley.edu/classes/es196/projects/2017final/FerriterM_2017.pdf

Ficke, A.D., Myrick, C.A., and Hansen, L.J. 2007. "Potential Impacts of Global Climate Change on Freshwater Fisheries." *Reviews in Fish Biology and Fisheries* 17 (4): 581–613. https://doi.org/10.1007/s11160-007-9059-5.

Figueroa J. 1834. *Diario de la expedicion al otro lado de la Bahia de San Francisco por el general Figueroa.* José Figueroa papers, 1833-1835, BANC MSS C-A 239. *Courtesy of The Bancroft Library, UC Berkeley.*

Fischer, J., and Lindenmayer, D.B. 2007. "Landscape Modification and Habitat Fragmentation: A Synthesis." *Global Ecology and Biogeography* 16 (3): 265–80. https://doi.org/10.1111/j.1466-8238.2007.00287.x.

Fitzgerald, R. 2013. "Summary of TMDL Development Data Pertaining to Nutrient Impairments in the Laguna de Santa Rosa Watershed [Revised]." Santa Rosa, CA: North Coast Regional Water Quality Control Board.

van Fleet, C.C. 1917. Egg card reference for *Agelaius phoeniceus*. Catalog #111516. *Courtesy of Western Foundation of Vertebrate Zoology, www.collections.wfvz.org.*

Flint, L.E., and Flint, A.L. 2012. "Simulation of Climate Change in San Francisco Bay Basins, California: Case Studies in the Russian River Valley and Santa Cruz Mountains." USGS Numbered Series 2012–5132. Scientific Investigations Report. Reston, VA: U.S. Geological Survey. http://pubs.er.usgs.gov/publication/sir20125132.

Flint, L.E., Flint, A.L., Mendoza, J., Kalansky, J., and Ralph, F. M. 2018. "Characterizing Drought in California: New Drought Indices and Scenario-Testing in Support of Resource Management." *Ecological Processes* 7 (1): 1. https://doi.org/10.1186/s13717-017-0112-6.

Folke, C., Holling, C.S., and Perrings, C. 1996. "Biological Diversity, Ecosystems, and the Human Scale." *Ecological Applications* 6 (4): 1018–24. https://doi.org/10.2307/2269584.

Folke, C., Carpenter, S., Walker, B., Scheffer, M., Elmqvist, T., Gunderson, L., and Holling, C.S. 2004. "Regime Shifts, Resilience, and Biodiversity in Ecosystem Management." *Annual Review of Ecology, Evolution, and Systematics* 35 (1): 557–81. https://doi.org/10.1146/annurev.ecolsys.35.021103.105711.

Fredrickson, D.A., and Markwyn, D.W. 1990. "Cultural Resources of the Laguna." In *History, Land Uses and Natural Resources of the Laguna de Santa Rosa*, David W. Smith Consulting, 3-1 to 3–12.

Fried, J.S., Torn, M.S., and Mills, E. 2004. "The Impact of Climate Change on Wildfire Severity: A Regional Forecast for Northern California." *Climatic Change* 64 (1): 169–91. https://doi.org/10.1023/B:CLIM.0000024667.89579.ed.

Gibbs, J.P., and Melvin, S. 1992. "American Bittern." In *Migratory Nongame Birds of Management Concern in the Northeast*, 51–88. Newton Corner, MA: U.S. Fish and Wildlife Service.

Gillard, M., Grewell, B.J., Deleu, C., and Thiébaut, G. Climate warming and water primroses: germination responses of populations from two invaded ranges. Aquatic Botany 136:155-163. DOI 10.1016/j.aquabot.2016.10.001. 2017.

Gray, N. 1857. *Field Notes of the Final Survey of the Rancho Los Molinos, John B.R. Cooper, Confirmee.* Vol. Book C-25/G3. U.S. Department of the Interior, Bureau of Land Management.

Gregory, T. 1911. *History of Sonoma County, California, with Biographical Sketches of the Leading Men and Women of the County, Who Have Been Identified with Its Growth and Development from the Early Days to the Present Time.* Los Angeles, CA: Hisoric Record Company.

Grewell, B.J., and Futrell, C.J. 2009. "Restoration and Management of Ludwigia Hexapetala-Invaded Wetlands of the Laguna in the Face of Climate Change." In USDA-Agricultural Research Service Exotic & Invasive Weed Research Unit.

Grewell, B.J., Skaer Thomason, M.J., Futrell, C.J., Iannucci, M., and Drenovsky, M.E. 2016a. "Trait Responses of Invasive Aquatic Macrophyte Congeners: Colonizing Diploid Outperforms Polyploid." *AoB PLANTS* 8. https://doi.org/10.1093/aobpla/plw014.

Grewell, B.J., Netherland, M.D., and Skaer-Thomason, M.J. 2016b. "Establishing Research and Management Priorities for Invasive Water Primroses (Ludwigia Spp.):" Fort Belvoir, VA: Defense Technical Information Center. https://doi.org/10.21236/AD1002917.

Grewell, B.J., Gillard, M.B., Futrell, C.J., and Castillo, J.M. 2019. "Seedling Emergence from Seed Banks in Ludwigia Hexapetala-Invaded Wetlands: Implications for Restoration." *Plants* 8 (11): 451. https://doi.org/10.3390/plants8110451.

Grismer, M.E., O'Geen, A.T., and Lewis, D. 2006. "Vegetation Filter Strips for Nonpoint Source Pollution Control in Agriculture." 8195. University of California Division of Agriculture and Natural Resources.

Grossinger, R.M., Striplen, C.J., Askevold, R.A., Brewster, E., and Beller, E.E. 2007. "Historical Landscape Ecology of an Urbanized California Valley: Wetlands and Woodlands in the Santa Clara Valley." *Landscape Ecology* 22 (S1): 103–20. https://doi.org/10.1007/s10980-007-9122-6.

Harpole, W.S., Ngai, J.T., Cleland, E.E., Seabloom, E.W., Borer, E.T., Bracken, M.E.S., Elser, J.J., et al. 2011. "Nutrient Co-Limitation of Primary Producer Communities." *Ecology Letters* 14 (9): 852–62. https://doi.org/10.1111/j.1461-0248.2011.01651.x.

He, M., and Gautam, M. 2016. "Variability and Trends in Precipitation, Temperature and Drought Indices in the State of California." *Hydrology* 3 (2): 14. https://doi.org/10.3390/hydrology3020014.

Healdsburg Enterprise. 1890. "From Pop. Two Miles from Forestville, March 19, 1890." March 26, 1890, 44 edition. *Courtesy of California Digital Newspaper Collection.*

———. 1926. "Reclaiming 1000 Acres of Land near Sebastopol." September 9, 1926. *Courtesy of California Digital Newspaper Collection.*

Healdsburg Tribune. 1927. "Sonoma County." November 1, 1927. *Courtesy of California Digital Newspaper Collection.*

Henning, J.A., Gresswell, R.E., and Fleming, I.A. 2006. *Juvenile Salmonid Use of Freshwater Emergent Wetlands in the Floodplain and Its Implications for Conservation Management.* Vol. 26.

Hilty, J.A., and Merenlender, A.M. 2004. "Use of Riparian Corridors and Vineyards by Mammalian Predators in Northern California." *Conservation Biology* 18 (1): 126–35. https://doi.org/10.1111/j.1523-1739.2004.00225.x.

Holmes, A.L, Humple, D.L., Gardali, T., and Geupel, G. 1999. "Songbird Habitat Associations and Response to Disturbance in the Point Reyes National Seashore and Golden Gate National Recreation Area." Stinson Beach, CA: Point Reyes Bird Observatory.

Holmes, L.C., and Nelson, J.W. 1914. "Reconnaissance Soil Survey of the San Francisco Bay Region, California." Washington, D.C. Government Printing Office.

———. 1915. "Reconnaissance Soil Survey of the Sacramento Valley, California." Washington, D.C: U.S. Department of Agriculture, Bureau of Soils. Government Printing Office.

Holway, R.S. 1913. "The Russian River: A Characteristic Stream of the California Coast Ranges." University of California Publications in Geography.

Honton, J., and Sears, A.W. 2006. *Enhancing and Caring for the Laguna.* Santa Rosa, CA: Laguna de Santa Rosa Foundation.

Isola, C.R., Colwell, M.A., Taft, O.W. and Safran, R.J., 2000. "Interspecific differences in habitat use of shorebirds and waterfowl foraging in managed wetlands of California's San Joaquin Valley." *Waterbirds*, pp.196-203.

IUCN (International Union for the Conservation of Nature). 2000. IUCN Guidelines for the Prevention of Biodiversity Loss caused by Alien Invasive Species (as approved by 51st Meeting of IUCN Council, February 2000). IUCN Information paper. Available at: http://www. issg.org/pdf/aliens_newsletters/supplementissue11.pdf

Jarvie, H.P., Sharpley, A.N., Withers, P.J.A., Scott, J.T., Haggard, B.E., and Neal, C. 2013. "Phosphorus Mitigation to Control River Eutrophication: Murky Waters, Inconvenient Truths, and 'Postnormal' Science." *Journal of Environmental Quality* 42 (2): 295–304. https://doi.org/10.2134/jeq2012.0085.

Kaushal, S.S., Groffman, P.M., Mayer, P.M. Striz, E., and Gold, A.J. 2008. "Effects of Stream Restoration on Denitrification in an Urbanizing Watershed." *Ecological Applications* 18 (3): 789–804. https://doi.org/10.1890/07-1159.1.

Keeley, J.E. 2001. "Fire and Invasive Species in Mediterranean-Climate Ecosystems of California." Pages 81–94 in K.E.M. Galley and T.P. Wilson (eds.). *Proceedings of the Invasive Species Workshop: the Role of Fire in the Control and Spread of Invasive Species.* Fire Conference 2000: the First National Congress on Fire Ecology, Prevention, and Management. Miscellaneous Publication No. 11, Tall Timbers Research Station, Tallahassee, FL..

Kieser & Associates, LLC. 2015. "Water Quality Trading Framework for the Laguna de Santa Rosa Watershed, California." Technical Report. Sonoma Resource Conservation District.

Krawchuk, Meg, and Max Moritz. 2012. "Fire and Climate Change in California: Changes in the Distribution and Frequency of Fire in Climates of the Future and Recent Past (1911-2099)." CEC-500-2012-026. Sacramento, CA: California Energy Commission.

Kroeber, A.L. 1925. *Handbook of the Indians of California.* New York, NY: Dover Publications, Inc.

Laguna de Santa Rosa Foundation. 2011. "Ecology of the Laguna." http://www.lagunafoundation.org/about_ecology.html

Laguna de Santa Rosa Foundation. 2016. "Restoration along the Middle Reach of the Laguna de Santa Rosa." *Meanderings*, W. Trowbridge.

Laguna TAC (Technical Advisory Committee). 1989. "Fish and Wildlife Restoration of the Laguna de Santa Rosa, Sonoma County, California."

Laurel Marcus & Associates. 2004. "Copeland Creek Watershed Assessment." Oakland, CA: Laurel Marcus & Associates.

Laymon, S.A., and Halterman, M.D. 1989. "A Proposed Habitat Management Plan for Yellow-Billed Cuckoos in California." General Technical Report PSW-110. USDA Forest Service.

Lee, C. 1944. "Sonoma County Sanitary Survey: Report on Sanitary Survey of Sonoma County, California with Recommendations for Control of Epidemics [Unpublished]." San Francisco, CA: Sonoma County Board of Supervisors and the Sonoma County Department of Public Health.

Lewis, W.M., Wurtsbaugh, W.A., and Paerl, H.W. 2011. "Rationale for Control of Anthropogenic Nitrogen and Phosphorus to Reduce Eutrophication of Inland Waters." *Environmental Science & Technology* 45 (24): 10300–305. https://doi.org/10.1021/es202401p.

Luber, G., and McGeehin, M. 2008. "Climate Change and Extreme Heat Events." *American Journal of Preventive Medicine*, Theme Issue: Climate Change and the Health of the Public, 35 (5): 429–35. https://doi.org/10.1016/j.amepre.2008.08.021.

Ludwigia Task Force. 2004. "Control of Ludwigia Hexapetala Infestations in Laguna and Wilfred/ Bellevue Flood Control Channels, and the Laguna Wildlife Area, Sonoma County, California." Santa Rosa, CA: Sonoma County Ludwigia Task Force.

Mann, M.L., Batllori, E., Moritz, M.A., Waller, E.K., Berck, P., Flint, A.L., Flint, L.E., and Dolfi, E. 2016. "Incorporating Anthropogenic Influences into Fire Probability Models: Effects of Human Activity and Climate Change on Fire Activity in California." *PLOS ONE* 11 (4): e0153589. https://doi.org/10.1371/journal.pone.0153589.

Marryat F. 1855. *Mountains and Molehills or Recollections of a Burnt Journal*. London: Longman, Brown, Green, and Longmans.

Martin, H.B. 1859. "Plat and Field Notes of Road from Santa Rosa to Healdsburg." Sonoma County Recorder.

Mayer, P.M., Reynolds, S.K., McCutchen, M.D., and Canfield, T.J. 2007. "Meta-Analysis of Nitrogen Removal in Riparian Buffers." *Journal of Environmental Quality* 36 (4): 1172–80. https://doi.org/10.2134/jeq2006.0462.

MEA (Millennium Ecosystem Assessment). 2005. *Ecosystems and Human Wellbeing: Synthesis*. Washington, D.C: Island Press.

Meisler, J. 2008. "Ludwigia Control Project." Santa Rosa, CA: Laguna de Santa Rosa Foundation.

———. 2009. "Lessons from Ludwigia Control in Sonoma County." *Cal-IPC News* 17 (2): 16.

Menefree, C.A. 1873. *Historical and Descriptive Sketch Book of Napa, Sonoma, Lake, and Mendocino: Comprising Sketches of Their Topography, Productions, History, Scenery, and Peculiar Attractions.* Berkeley, CA: California Indian Library Collections.

Merritt Smith Consulting. 1995. "Santa Rosa Subregional Long-Term Wastewater Project, Anadromous Fish Migration Study." Prepared for Harland Bartholomew Associates and the City of Santa Rosa. April.

Meyer, H. 1877. "The Phantom of the Laguna." *Sonoma Democrat*, March 31, 1877, 23 edition. *Courtesy of California Digital Newspaper Collection.*

Micheli, E.R., Flint, L.E., and Veloz, S. 2009. "Climate and Hydrology Summary Report: Laguna de Santa Rosa." Petaluma, CA: Point Blue Conservation Science.

Micheli, E.R., Flint, L.E., Veloz, S., Johnson, K., and Heller, N.E. 2016. "Climate Ready North Bay: Vulnerability Assessment Data Products." Technical Memorandum. Santa Rosa, CA: Pepperwood Preserve.

Miller, G.W. 1960. *Reconnaissance Study for Laguna de Santa Rosa Pilot Channel, Trenton Bridge to Sebastopol Road.* Sonoma County Flood Control and Water Conservation District.

Millington, S. 1865. *Transcript of the Field Notes of the Survey of the Subdivision Lines Township 7 North Range 9 West, Mount Diablo Meridian, State of California.* Vol. Book 182-9.

Mitsch, W.J., and Gosselink, J.C. 1993. *Wetlands.* 2nd ed. New York, NY: Van Nostrand Reinhold.

Moody, J.A., and Martin, D.A. 2009. "Synthesis of Sediment Yields after Wildland Fire in Different Rainfall Regimes in the Western United States." *International Journal of Wildland Fire* 18 (1): 96–115. https://doi.org/10.1071/WF07162.

Moraga, G. 1810. *Diario de Su Experdicion al Puerto de Bodega. Provincial State Papers Tom. XIX 1805-1815.* BANC MSS C-A 12. https://ia802607.us.archive.org/5/items/168036075_79_14/168036075_79_14.pdf.

Morris, C.N. 1995. "Waste Reduction Strategy for the Laguna de Santa Rosa." North Coast Regional Water Quality Control Board.

Moritz, M.A., Hessburg, P.F. and Povak, N.A., 2011. Native fire regimes and landscape resilience. In The landscape ecology of fire (pp. 51-86). Springer, Dordrecht. Dietz, Michael E., and John C. Clausen. 2008. "Stormwater Runoff and Export Changes with Development in a Traditional and Low Impact Subdivision." *Journal of Environmental Management*, Microbial and Nutrient Contaminants of Fresh and Coastal Waters, 87 (4): 560–66. https://doi.org/10.1016/j.jenvman.2007.03.026.

Mount, J., Gray, B., Chappelle, C., Gartrell, G., Grantham, T., Moyle, P., Seavy, N.E., Szeptycki, L., and Thompson, B.B. 2017. "Managing California's Freshwater Ecosystems: Lessons from the 2012–16 Drought." Public Policy Institute of California.

Moyle, P.B. 2002. *Inland Fishes of California.* 1st ed. Berkeley, CA: University of California Press.

NAIP (National Aerial Imagery Program). 2016. "Modern Aerial Ortho-Imagery." USDA Farm Service Agency.

NCRWQCB (North Coast Regional Water Quality Control Board). 2018. *Water Qualtiy Trading Framework for the Laguna de Santa Rosa Watershed.*

Nishikawa, T., Hevesi, J., Sweetkind, D., and Woolfenden, L. 2013. "Hydrologic and Geochemical Characterization of the Santa Rosa Plain Watershed, Sonoma County, California." Scientific Investigations Report 2013–5118. Reston, VA: US Geological Survey.

Nixon, S.W. 1995. "Coastal Marine Eutrophication: A Definition, Social Causes, and Future Concerns." *Ophelia* 41 (1): 199–219. https://doi.org/10.1080/00785236.1995.10422044.

NMFS (National Marine Fisheries Service). 2010. "Recovery Plan for the Evolutionarily Significant Unit of Central California Coast Coho Salmon." Santa Rosa, CA: National Marine Fisheries Service, Southwest Region.

Okada, M., Grewell, B.J. and Jasieniuk, M. 2009. "Clonal Spread of Invasive Ludwigia Hexapetala and L. Grandiflora in Freshwater Wetlands of California." *Aquatic Botany* 91 (3): 123–29. https://doi.org/10.1016/j.aquabot.2009.03.006.

Opperman, J.J, Moyle, P.B., Larsen, E.W., Florsheim, J.L., and Manfree, A.D. 2017. *Floodplains: Processes and Management for Ecosystem Services.* Oakland, CA: University of California Press.

Origer, R.M., and FredricksonD.A. 1980. *The Laguna Archaeological Research Project, Sonoma County, California.* Cultural Resources Facility, Anthropological Studies Center, Sonoma State University.

Pacific Rural Press. 1880. "Agricultural Notes," April 17, 1880. Courtesy of California Digital Newspaper Collection.

Paerl, H.W. 2009. Controlling Eutrophication along the Freshwater–Marine Continuum: Dual Nutrient (N and P) Reductions are Essential. Estuaries and Coasts 32, 593–601 (2009). https://doi.org/10.1007/s12237-009-9158-8

Parker, I.M.,Simberloff, D., Lonsdale, W.M., Goodell, K., Wonham, M., Kareiva, P.M., Williamson, M.H., et al. 1999. "Impact: Toward a Framework for Understanding the Ecological Effects of Invaders." *Biological Invasions* 1 (1): 3–19. https://doi.org/10.1023/A:1010034312781.

Petaluma Courier. 1883. "News from Our Neighbors." January 24, 1883. *Courtesy of Newspapers.com.*

Petaluma Weekly Argus. 1881. "Sonoma County Items." January 7, 1881. *Courtesy of Newspapers.com.*

Petaluma Weekly Argus. ———. 1885. "Cat-Fish Hunters." June 27, 1885. *Courtesy of Newspapers.com.*

Peter, S., Rechsteiner, R., Lehmann, M.F., Brankatschk, R., Vogt, T., Diem, S., Wehrli, B., Tockner, K., and Durisch-Kaiser, E. 2012. "Nitrate Removal in a Restored Riparian Groundwater System: Functioning and Importance of Individual Riparian Zones." *Biogeosciences* 9 (11): 4295–4307. https://doi.org/10.5194/bg-9-4295-2012.

Pierce, D.W., Cayan, D.R., Das, T., Maurer, E.P., Miller, N.L., Bao, Y., Kanamitsu, M., et al. 2013a. "The Key Role of Heavy Precipitation Events in Climate Model Disagreements of Future Annual Precipitation Changes in California." *Journal of Climate* 26 (16): 5879–96. https://doi.org/10.1175/JCLI-D-12-00766.1.

Pierce, D.W., Cayan, D.R., Das, T., Maurer, E.P., Miller, N.L., Bao, Y., Kanamitsu, M., et al. 2013b. "Probabilistic Estimates of Future Changes in California Temperature and Precipitation Using Statistical and Dynamical Downscaling." *Climate Dynamics* 40 (3): 839–56. https://doi.org/10.1007/s00382-012-1337-9.

Pierce, D. W., Kalansky, J.F., and Cayan, D.R. 2018. Climate, Drought, and Sea Level Rise Scenarios for the Fourth California Climate Assessment. California's Fourth Climate Change Assessment, California Energy Commission. Publication Number: CNRA-CEC-2018-006.

Pimentel, D., Zuniga, R. and Morrison, D. 2005. "Update on the Environmental and Economic Costs Associated with Alien-Invasive Species in the United States." *Ecological Economics*, Integrating Ecology and Economics in Control Bioinvasions, 52 (3): 273–88. https://doi.org/10.1016/j.ecolecon.2004.10.002.

Potter, C., and Hiatt, C. 2009. "Modeling River Flows and Sediment Dynamics for the Laguna de Santa Rosa Watershed in Northern California." *Journal of Soil and Water Conservation* 64 (6): 383–93. https://doi.org/10.2489/jswc.64.6.383.

Press Democrat. 1886. "Local Brevities." February 23, 1886. *Courtesy of California Digital Newspaper Collection.*

PRISM Climate Group. 2019. "PRISM Home Page." Northwest Alliance for Computational Science and Engineering. http://www.prism.oregonstate.edu/.

PWA (Philip Williams & Associates, Ltd.). 2004a. *Laguna de Santa Rosa Feasibility Study: Year One Geomorphic Investigation: Final Report*. Vol. Ref. #1411-08.

———. 2004b. *Sediment Sources, Rate and Fate in the Laguna de Santa Rosa, Sonoma County, CA: Final Report, Volume 2*. Vol. Ref. #1411-20.

Pyke, C. Warren, M.P., Johnson, T., LaGro, J., Scharfenberg, J., Groth, P., Freed, R. William Schroeer, and Main, E. 2011. "Assessment of Low Impact Development for Managing Stormwater with Changing Precipitation Due to Climate Change." *Landscape and Urban Planning* 103 (2): 166–73. https://doi.org/10.1016/j.landurbplan.2011.07.006.

Reese, D.A.. 1996. "Comparative Demography and Habitat Use of Western Pond Turtles." Berkeley, CA: University of California, Berkeley.

Richmond, O.M., Hines, J.E. and Beissinger, S.R. 2010. "Two-Species Occupancy Models: A New Parameterization Applied to Co-Occurrence of Secretive Rails." *Ecological Applications* 20 (7): 2036–46. https://doi.org/10.1890/09-0470.1.

Riffell, S.K., Keas, B.E., and Burton, T.M. 2001. "Area and Habitat Relationships of Birds in Great Lakes Coastal Wet Meadows." *Wetlands* 21 (4): 492–507. https://doi.org/10.1672/0277-5212(2001)021[0492:AAHROB]2.0.CO;2.

Robbins, G.T. 1937. [Record for *Ranunculus lobbii*.] Field Notebook #1, 1937-1944. Jepson Herbarium, University of California, Berkeley.

Ross, Sarepta Ann Turner. 1914. "Mary Elizabeth Turner: Recollections of Pioneer Days Written by a Pioneer." *The Republican*, December 1914.

Rubtzoff, P. 1964. "Notes on Fresh-water Marsh and Aquatic Plants in California - III." *Leaflets of Western Botany* 10 (15): 306-312.

———. 1966. "Notes on Fresh-water Marsh and Aquatic Plants in California - VI." *Leaflets of Western Botany* 10 (5): 68-73.

Safford, H.D., van de Water, K., and Clark, C. 2013. California Fire Return Interval Departure (FRID) map, 2012 version. USDA Forest Service, Pacific Southwest Region, Sacramento and Vallejo, CA. http://www.fs.usda.gov/main/r5/landmanagement/gis.

SCACO (Sonoma County Agricultural Commissioner's Office). 2013. "Best Management Practices for Agricultural Erosion and Sediment Control," December, 45.

SCEDB. 2018. "Sonoma County City Profile and Projections Report." Sonoma County Economic Development Board.

Schindler, D.W., Hecky, R.E., Findlay, D.L., Stainton, M.P., Parker B.R., Paterson, M., Beaty, K.G., Lyng, M. and Kasian, S.E.M. 2008. Eutrophication of lakes cannot be controlled by reducing nitrogen input: Results of a 37 year whole ecosystem experiment. Proceedings of the National Academy of Science USA 105: 11254–11258. doi:10.1073/pnas.0805108105.

Schipper, L.A., Cameron, S.C. and Warneke, S. 2010. "Nitrate Removal from Three Different Effluents Using Large-Scale Denitrification Beds." *Ecological Engineering*, Managing Denitrification in Human Dominated Landscapes, 36 (11): 1552–57. https://doi.org/10.1016/j.ecoleng.2010.02.007.

SCWA (Sonoma County Water Agency), City of Santa Rosa, City of Sebastopol, CDFW (California Department of Fish and Game), Sonoma County Agricultural Preservation and Open Space District, and Laguna de Santa Rosa Foundation. 2016. "Restoration Planning in the Laguna de Santa Rosa: A Science-Based Partnership between Non-Profit Agencies and Public and Private Land Owners." Poster.

Seavy, N.E., Gardali, T., Golet, G.H., Griggs, F.T., Howell, C.A., Kelsey, R., Small, S.L., Viers, J.H. and Weigand, J.F., 2009. Why climate change makes riparian restoration more important than ever: recommendations for practice and research. Ecological Restoration, 27(3), pp.330-338.

Semlitsch, R.D., and Bodie, J.R. 2003. "Biological Criteria for Buffer Zones around Wetlands and Riparian Habitats for Amphibians and Reptiles." *Conservation Biology* 17 (5): 1219–28. https://doi.org/10.1046/j.1523-1739.2003.02177.x.

SFEI-ASC (San Francisco Estuary Institute-Aquatic Science Center). 2014. "NCARI: North Coast Aquatic Resource Inventory Mapping."

SFEI (San Francisco Estuary Institute) 2017a. "Demonstration of a Watershed Approach To Wetland Restoration Planning Using GreenPlan-IT." Technical Memorandum. Richmond, CA: San Francisco Estuary Institute.

——— 2017b. Simulating Wetland Assimilation Capacity in the Lower Santa Rosa Creek Watershed Pilot Demonstration Area Using the HSPF Model. Technical memo prepared by the San Francisco Estuary Institute-Aquatic Science Center, Richmond, CA. SWRCB STD Agreement Number: 15-021-250-1.

Shelton, Alfred C. 1911. "Nesting of the California Cuckoo." *The Condor* 13 (1): 19–22.

Short, F.T., Kosten, S., Morgan, P.A., Malone, S. and Moore, G.E., 2016. Impacts of climate change on submerged and emergent wetland plants. Aquatic Botany, 135, pp.3-17.

Skaer Thomason, M.J., Grewell, B.J., and Netherland. M.D., 2018a. "Dynamics of Ludwigia Hexapetala Invasion at Three Spatial Scales in a Regulated River." *Wetlands* 38 (6): 1285–98. https://doi.org/10.1007/s13157-018-1053-2.

Skaer Thomason, M.J., McCort, C.D., Netherland, M.D., and Grewell, B.J. 2018b. "Temporal and Nonlinear Dispersal Patterns of Ludwigia Hexapetala in a Regulated River." *Wetlands Ecology and Management* 26 (5): 751–62. https://doi.org/10.1007/s11273-018-9605-z.

Sloop, C., Honton, J., Creager, C., Chen, L., Andrews, E.S., and Bozkurt, S. 2007. *The Altered Laguna: A Conceptual Model for Watershed Stewardship*. Santa Rosa, CA: Laguna de Santa Rosa Foundation.

Sloop, C. 2009. "Ramsar Designation of the Laguna de Santa Rosa Wetland Complex as a Wetland of International Significance." SOL Poster.

Sloop, C., and Hug, L. 2009. "Bird Inventory and Monitoring at Laguna de Santa Rosa: Years 2004/05 & 2007-2009." 395. Santa Rosa, CA: Laguna de Santa Rosa Foundation.

Sloop, C., and Jones, W. 2010. "Information Sheet on Ramsar Wetlands." https://rsis.ramsar.org/RISapp/files/RISrep/US1930RIS.pdf.

Smith, D.W. 1990. "History, Land Uses, and Natural Resources of the Laguna de Santa Rosa." Unpublished Report. Santa Rosa Subregional Water Reclamation System.

Smith, V.H. 2003. "Eutrophication of Freshwater and Coastal Marine Ecosystems a Global Problem." *Environmental Science and Pollution Research* 10 (2): 126–39. https://doi.org/10.1065/espr2002.12.142.

Solé, R.V., Levin, S.A., Dent, C.L., Cumming, G.S., and Carpenter. S.R. 2002. "Multiple States in River and Lake Ecosystems." *Philosophical Transactions of the Royal Society of London. Series B: Biological Sciences* 357 (1421): 635–45. https://doi.org/10.1098/rstb.2001.0991.

Sonoma County Democrat. 1861. Grizzly killed. May 28. *Courtesy of Sonoma State University Library*.

Sonoma Democrat. 1875. "Fish." January 2, 1875.

Sonoma Democrat. 1879. "The Best Part of the County." June 7, 1879.

Sonoma Democrat. 1882. "Local Notes." February 18, 1882.

Sonoma Veg Map. 2017. "Sonoma County Fine Scale Vegetation and Habitat Map." Sonoma County Water Agency, Sonoma County Agricultural Preservation and Open Space District, Sonoma County Vegetation Mapping and LiDAR Program. http://sonomavegmap.org/.

Spence, B.C., Harris, S.L., Jones, W.E., Goslin, M.N., Agrawal, A., and Mora, E. 2005. "Historical Occurrence of Coho Salmon in Streams of the Central California Coast Coho Salmon Evolutionarily Significant Unit." NOAA Technical Memorandum NMFS-SWFSC-383. U.S. Department of Commerce, National Oceanic and Atmospheric Adminstration.

Stewart, O.. 1943. "Notes on Pomo Ethnogeography." University of California Publications in American Archaeology and Ethnology vol 40, no. 2, pp. 29-62. Berkeley, CA: University of California Press.

Sutula, M.R., Mazor, R., and Theroux, S. 2018. "Scientific Bases for Assessment, Prevention, and Management of Biostimulatory Impacts in California Wadeable Streams." Technical Report 1048. Costa Mesa, CA: Southern California Coastal Water Research Project.

Sweeney, B.W., and Newbold J.D. 2014. "Streamside Forest Buffer Width Needed to Protect Stream Water Quality, Habitat, and Organisms: A Literature Review." *JAWRA Journal of the American Water Resources Association* 50 (3): 560–84. https://doi.org/10.1111/jawr.12203.

Tarnay, L. 2018. Living with Fire (and Smoke) in California. Presentation given at the Living with Fire in California's Coast Ranges Symposium. May 7-9, 2018. Sonoma State University. https://www.pepperwoodpreserve.org/wp-content/uploads/2018/08/2018-Living-with-Fire-9-Smoke-Tarnay.pdf

Taylor, B. 1862. *At Home and Abroad: A Sketch-Book of Life, Scenery and Men*. New York, NY: G. P. Putnam.

Taylor, S.L., Roberts, SC., Walsh, C.J., and Hatt, B.E. 2004. "Catchment Urbanisation and Increased Benthic Algal Biomass in Streams: Linking Mechanisms to Management." *Freshwater Biology* 49 (6): 835–51. https://doi.org/10.1111/j.1365-2427.2004.01225.x.

TBCCC (Terrestrial Biodiversity Climate Change Collaborative). 2016. "Climate Ready Vegetation Report: Russian River Landscape Unit." Santa Rosa, CA: Pepperwood Preserve.

Teels, B.M., Rewa, C.A., and Myers,J. 2006. "Aquatic Condition Response to Riparian Buffer Establishment." *Wildlife Society Bulletin* 34 (4): 927–35. https://doi.org/10.2193/0091-7648(2006)34[927:ACRTRB]2.0.CO;2.

Tetra Tech. 2015a. "Laguna de Santa Rosa Nutrient Analysis (Revised)." Research Triangle Park, NC: Tetra Tech, Inc.

———. 2015b. "Laguna de Santa Rosa Sediment Budget." Research Triangle Park, NC: Tetra Tech, Inc.

Tetra Tech 2020. "Laguna de Santa Rosa – Linkage Analysis for Sediment Impairments (Revised)". Research Triangle Park, NC: Tetra Tech, Inc. Jaunary 10.

The Press Democrat. 1920. "Big Ditch Dug at Half Price Asked in Bids," December 16, 1920.

———. 1928. "Ballard Lake Falls; Large Area Now Dry," September 20, 1928. *Courtesy of Newspapers.com*.

The Sebastopol Times. 1903a. "Metropolis of the Thrifty Gold Ridge - Sebastopol Making Vigorous Growth." January 2, 1903. *courtesy Sonoma State University Library, North Bay Digital Collections.*

———. 1903b. "The Best Part of the County." January 2, 1903. *Courtesy of Sonoma State University Library, North Bay Digital Collections.*

Thouvenot, L., Haury, J., and Thiébaut, G. 2013. A success story: water primroses, aquatic plant pests. Aquatic Conservation: Marine and Freshwater Ecosystems 23:790-803.

Tomasko, D.A., Britt, M. and Carnevale, M.J. 2016. "The Ability of Barley Straw, Cypress Leaves and L-lysine to Inhibit Cyanobacteria in Lake Hancock, a Hypereutrophic Lake in Florida." *Florida Scientist* 79 (2-3): 147-158.

Torrey, J., Bigelow, J.M., and Engelmann, G. 1857. "Report on the Botany of the Expedition." Reports of Explorations and Surveys for a Railroad Route from the Mississippi River to the Pacific Ocean. Washington, D.C.: War Department.

Tracy, C.C. 1859a. *Field Notes of the Obsolete Survey of the Rancho Llano de Santa Rosa, Joaquin Carillo, Confirmee.* Vol. Book G10. U.S. Department of the Interior, Bureau of Land Management.

———. 1859b. "Plat of the Llano de Santa Rosa, Finally Confirmed to Joaquin Carillo. Surveyed under Instructions from the U.S. Surveyor General. Land Case Map E-131."

Turbelin, A.J., Malamud, B.D., and Francis, R.A. 2017. "Mapping the Global State of Invasive Alien Species: Patterns of Invasion and Policy Responses." *Global Ecology and Biogeography* 26 (1): 78–92. https://doi.org/10.1111/geb.12517.

Unknown. 1870. "Map of the Cotate Rancho." Curtis and Associates, Inc.

U.S. Surveyor General's Office. 1865. "Map of Township No. 7 North Range 8 West (Mount Diablo Meridian)." San Francisco, CA. Bureau of Land Management.

———. 1868. "Map of Township No. 8 North Range 9 West (Mount Diablo Meridian)." San Francisco, CA. Bureau of Land Management.

USDA (U. S. Department of Agriculture). 1942. "[Aerial Photos of Sonoma County] Flight COF-1942." U.S. Department of Agriculture, Soil Conservation Service.

USDC (U.S. District Court, Northern District). 1840. "Mapa Bel [Sic] Rancho St. Micuel [Sic], 3 Leguas : [Sonoma County, Calif.]. Land Case Map B-664." Bancroft Library.

———. 1849. "San Miguel de Marcos West, 6 Leguas: [Sonoma County, Calif.]. Land Case Map D-664a." Bancroft Library.

USFWS (US Fish and Wildlife Service). 2005. "Santa Rosa Plain Conservation Strategy." Sacramento, CA: US Fish and Wildlife Service.

———. 2011. "California Freshwater Shrimp (Syncaris Pacifica) 5-Year Review: Summary and Evaluation." Sacramento, CA: US Fish and Wildlife Service.

———. 2016. "Recovery Plan for the Santa Rosa Plain." Sacramento, CA: US Fish and Wildlife Service.

Vallejo M.G., Farris, G., and Beebe, R. 2000. *Report of a Visit to Fort Ross and Bodega Bay*. California Mission Studies Association, Occassional Paper #4. http://www.fortross.org/lib/46/report-of-a-visitto-fort-ross-and-bodega-bay-in-april-1833-by-mariano-g-vallejo.pdf. Accessed March 13, 2017. *Courtesy of Fort Ross Conservancy Library*.

Verhoeven, J.T.A., Arheimer, B. Yin, C. and Hefting, M.M. 2006. "Regional and Global Concerns over Wetlands and Water Quality." *Trends in Ecology & Evolution* 21 (2): 96–103. https://doi.org/10.1016/j.tree.2005.11.015.

Vilà, M., Espinar J.L., Hejda, M., Hulme, P.E., Jarošík, V., Maron, J.L., Pergl, J., Schaffner, U., Sun, Y., and Pyšek, P. 2011. "Ecological Impacts of Invasive Alien Plants: A Meta-Analysis of Their Effects on Species, Communities and Ecosystems." *Ecology Letters*, 702–8. https://doi.org/10.1111/j.1461-0248.2011.01628.x@10.1111/(ISSN)1461-0248.oceans-to-mountains.

Vitousek, P.M, D'Antonio, C.M., Loope, L.L., Rejmánek, M., and Westbrooks, R. 1997. "Introduced Species: A Signficant Component of Human-Caused Global Change." *NEW ZEALAND JOURNAL OF ECOLOGY* 21 (1): 16.

Waaland, M. 1989. "Baseline Evaluation of Laguna de Santa Rosa Wetlands and Natural Resources." Technical Memorandum W1. Long-Term Detaile Wastewater Reclamation Studies. Golden Bear Biostudies.

Walker, W.S. 1880. *Gilmpses of Hungryland, or Califoria Sketches: Comprising Sentimental and Humorous Sketches, Poems, Etc., a Journey to California and Back Again, by Land and Water; Incidents of Every-Day Life on the Pacific Coast, Why I Came, What I Saw, and How I like It.* Cloverdale, CA: Reveille Publishing House.

Watson, E.B., Dean, D., Zinn, C.J., and Pendleton, R.L. 1915. "Soil Map, California. Healdsburg Sheet." U.S. Department of Agriculture, Bureau of Soils. Alabama Maps, University of Alabama.

Watson, E.B., Dean, D., Zinn, C.J., and Pendleton, R.L. 1917. "Soil Survey of the Healdsburg Area, California. Advance Sheets, Field Operations of the Bureau of Soils." U.S. Department of Agriculture, Bureau of Soils. Government Printing Office.

Welch, J.R. 2013. Sprouting Valley: Historical Ethnobotany of the Northern Pomo from Potter Valley, California. Society of Ethnobiology. Department of Geography, University of North Texas. Denton, TX.

Wells, G. 1920a. Egg card reference for *Cistothorus palustris*. Catalog #100971. *Courtesy of Western Foundation of Vertebrate Zoology, www.collections.wfvz.org*.

———. 1920b. Egg card reference for *Geothlypis trichas*. Catalog #109009. *Courtesy of Western Foundation of Vertebrate Zoology, www.collections.wfvz.org*.

———. 1923a. Egg card reference for *Catharus ustulatus*. Catalog #103333. *Courtesy of Western Foundation of Vertebrate Zoology, www.collections.wfvz.org*.

———. 1923b. Egg card reference for *Rallus limicola*. Catalog #44351. *Courtesy of Western Foundation of Vertebrate Zoology, www.collections.wfvz.org*.

———. 1926. Egg card reference for *Cardellina pusilla*. Catalog #108881. *Courtesy of Western Foundation of Vertebrate Zoology, www.collections.wfvz.org*.

Wenger, S., and Fowler, L. 2000. *Protecting Stream and River Corridors: Creating Effective Local Riparian Buffer Ordinances.* Public Policy Research Series. Athens, GA: Carl Vinson Institute of Government, the University of Georgia.

Werner, R.H., Hampson, R.P., Flaherty, J.M., Fagan, J.M., and Origer, T. 2003. Archaeological Investigations CA-SON-1019 Santa Rosa, Sonoma County, California. ASI Archaeology and Cultural Resources Management and Associates. Stockton, CA.

Wheatcroft, R.A., and Sommerfield, C.K. 2005. "River Sediment Flux and Shelf Sediment Accumulation Rates on the Pacific Northwest Margin." *Continental Shelf Research* 25 (3): 311–32. https://doi.org/10.1016/j.csr.2004.10.001.

Whitacre, T.H. 1853. *Copy of Field Notes of Survey of Township 6, 7, 8, 9, and 10 North of Ranges Numbers 6, 7, 8, 9, and 10 West of Mount Diablo Meridian in the State of California.* Vol. Book 210-5. U.S. Department of the Interior, Bureau of Land Management.

Wiess, S., Flint, A., Flint, L., Hamilton, H., Fernández, M., and Micheli, L. 2013. "High Resolution Climate-Hydrology Scenarios for San Francisco's Bay Area." Santa Rosa, CA: Pepperwood Preserve.

Winzler & Kelly-GHD. 2012. *Laguna-Mark West Creek Watershed Planning Scoping Study: Screening Technical Memorandum.*

Woodhouse, C.A., and G.T. Pederson. 2018. "Investigating Runoff Efficiency in Upper Colorado River Streamflow Over Past Centuries." *Water Resources Research* 54 (1): 286–300. https://doi.org/10.1002/2017WR021663.

Woolfenden, L.R., and Nishikawa, T. 2014. "Simulation of Groundwater and Surface-Water Resources of the Santa Rosa Plain Watershed, Sonoma County, California." USGS Numbered Series 2014–5052. Scientific Investigations Report. Reston, VA: U.S. Geological Survey. http://pubs.er.usgs.gov/publication/sir20145052.

von Wrangell, F.P., Stross, F., and Heizer, R.F. 1974. *Ethnographic Observations on the Coast Miwok and Pomo by Contre-Admiral F. P. Von Wrangell and P. Kostromitonov of the Russian Colony Ross, 1839.* Archaeological Research Facility, Department of Anthropology, University of California, Berkeley.

WSI (Watershed Sciences, Inc.). 2013. "[3-Ft. LiDAR-Derived Bare Earth Digital Elevation Model (DEM) for Sonoma County.]." Sonoma County Vegetation Mapping and LiDAR Consortium.

§

Appendix A

List of Additional California Native Plants Designated as Species of Local Concern

Common Name	Taxon Name	CNPS Ranking
Rincon Ridge manzanita	*Arctostaphylos stanfordiana ssp. decumbens*	1B.1
swamp harebell	*Campanula californica*	1B.2
Rincon Ridge ceanothus	*Ceanothus confusus*	1B.1
Vine Hill ceanothus	*Ceanothus foliosus var. vineatus*	1B.1
Peruvian dodder	*Cuscuta obtusiflora var. glandulosa*	2B.2
Golden larkspur	*Delphinium luteum*	1B.1
White seaside tarplant	*Hemizonia congesta ssp. congesta*	1B.2
Thin-lobed horkelia	*Horkelia tenuiloba*	1B.2
Baker's goldfields	*Lasthenia californica ssp. bakeri*	1B.2
Sebastopol meadowfoam	*Limnanthes vinculans*	1B.1
Marsh microseris	*Microseris paludosa*	1B.2
Saline clover	*Trifolium hydrophilum*	1B.2
Oval-leaved viburnum	*Viburnum ellipticum*	2B.3

California Native Plant Society (CNPS) Rankings

1B.1 = Plants that are rare, majority are endemic to California; is seriously threatened in California

1B.2 = Plants that are rare, majority are endemic to California; is moderately threatened in California

2B.1 = Plants that are rare in California, but common outside of California; seriously threatened in California

2B.2 = Plants that are rare in California, but common outside of California; moderately threatened in California

2B.3 = Plants that are rare in California, but common outside of California; not very threatened in California

Appendix B

List of California Native Wetland Plants Found in the Historical Record

The following table lists native wetland plants from the Laguna de Santa Rosa and its immediate surroundings with historical (pre-1950) voucher specimens recorded in the Consortium of California Herbaria (CCH, data provided by the participants of the Consortium of California Herbaria (ucjeps.berkeley.edu/consortium/)). The list was generated by querying the CCH database for records within the Laguna study area, then comparing the results against the Arid West 2016 Regional Wetland Plant List (Lichvar et al. 2016). Modern synonyms have been substituted for older botanical names. Where multiple records existed for the same species, the earliest recorded date and location are reported. Exotic species are excluded from this list. This table represents a partial record of historical plant species composition for wetland habitats in the Laguna; it does not constitute a comprehensive botanical inventory or restoration planting palette for the Laguna.

Family	Taxon Name	Common Name	Wetland Status*	Locality	Collection Date
Alismataceae	*Alisma triviale*	Northern Water-Plantain	OBL	2 mi NE Graton; Sebastopol Quad	5/17/1936
Alismataceae	*Damasonium californicum*	Fringed-Water-Plantain	OBL	W. of Santa Rosa (on rd. to Sebastopol, The Laguna)	7/29/1949
Alismataceae	*Sagittaria cuneata*	Arum-Leaf Arrowhead	OBL	Laguna near Fulton	6/1/1901
Alismataceae	*Sagittaria latifolia*	Duck-Potato	OBL	On road to Sebastopol the Laguna; marshy area W Santa Rosa; Laguna Rosa; Laguna marshy area	6/29/1949
Apiaceae	*Eryngium aristulatum*	California Eryngo	OBL	Along railroad tracks all pls. now dormant, W right School on Sebastopol road	1/1/1935

Family	Taxon Name	Common Name	Wet-land Status*	Locality	Collection Date
Apiaceae	*Eryngium armatum*	Coastal Eryngo	FACW	2 mi. SE Forestville; Sebastopol Quad	8/22/1937
Aristochloaceae	*Asarum caudatum*	Long-Tail Wild Ginger	FAC	Near Sebastopol	3/5/1947
Asteraceae	*Achillea millefolium*	Common Yarrow	FACU	Sebastopol	7/12/1907
Asteraceae	*Bidens frondosa*	Devil's-Pitchfork	FACW	The Laguna at Crossing of the Occidental Rd.	8/17/1946
Asteraceae	*Blennosperma nanum*	Common Stickyseed	FACW	Near Windsor	3/14/1902
Asteraceae	*Cirsium douglasii*	Douglas' Thistle	OBL	E of Santa Rosa; Santa Rosa Creek	7/2/1902
Asteraceae	*Grindelia hirsutula*	Hairy Gumweed	FACW	Eastern margin of Laguna. Upland. Along Occidental Rd. near bridge	7/30/1946
Asteraceae	*Helenium puberulum*	Rosilla	FACW	At Sebastopol	7/1/1902
Asteraceae	*Holozonia filipes*	Whitecrown	FACU	Lagoon at Sebastopol	7/1/1902
Asteraceae	*Lasthenia californica*	California Goldfields	FACU	Santa Rosa Creek	4/15/1902
Asteraceae	*Lasthenia fremontii*	Fremont's Goldfields	OBL	1 mile N of Windsor	5/10/1943
Asteraceae	*Layia chrysanthemoides*	Smooth Tidytips	FACW	Santa Rosa Valley. [noted on sheet: in meadow at Mark West, Sonoma County]	4/18/1864
Asteraceae	*Pseudognaphalium stramineum*	Cotton-Batting-Plant	FAC	Sebastopol	7/12/1907
Asteraceae	*Symphyotrichum chilense*	Pacific American-Aster	FAC	Sebastopol	8/1/1907
Asteraceae	*Symphyotrichum spathulatum*	Mountain American-Aster	FAC	Sebastopol	8/1/1907

Family	Taxon Name	Common Name	Wetland Status*	Locality	Collection Date
Asteraceae	*Wyethia angustifolia*	California-Compass-plant	FACU	Wright School on Sebastopol Road.	5/13/1935
Asteraceae	*Xanthium strumarium*	Rough Cocklebur	FAC	The Laguna north of Sebastopol	8/17/1946
Boraginaceae	*Plagiobothrys nothofulvus*	Rusty Popcorn-Flower	FAC	Santa Rosa Creek.	3/26/1902
Boraginaceae	*Plagiobothrys reticulatus*	Netted Popcorn-Flower	FACW	Sebastopol	6/6/1937
Boraginaceae	*Plagiobothrys undulatus*	Wavy-Stem Popcorn-Flower	OBL	Sebastopol	6/6/1937
Campanulaceae	*Downingia concolor*	Maroon-Spot Calico-Flower	OBL	Windsor	6/2/1921
Campanulaceae	*Githopsis specularioides*	Common Bluecup	FACU	North of Sebastopol	5/17/1880
Caprifoliaceae	*Symphoricarpos albus*	Common Snowberry	FACU	Sebastopol	7/12/1907
Caryophyllaceae	*Minuartia californica*	California Stitchwort	FACU	Near Windsor	2/14/1902
Celastraceae	*Euonymus occidentalis*	Western Wahoo	FACW	Along Jonive Creek about 1/8 mile NE of Wagnon Ranch, 5 mile W of Sebastopol	5/1/1939
Cyperaceae	*Carex athrostachya*	Slender-Beak Sedge	FACW	Sebastopol	6/6/1937
Cyperaceae	*Carex barbarae*	Santa Barbara Sedge	FAC	Sebastopol	5/17/1880
Cyperaceae	*Carex densa*	Dense Sedge	OBL	Sebastopol	6/6/1937
Cyperaceae	*Carex exsiccata*	Western Inflated Sedge	OBL	The Laguna north of Sebastopol	8/17/1946
Cyperaceae	*Carex pachystachya*	Thick-Head Sedge	FAC	Lagoon at Sebastopol	7/1/1902
Cyperaceae	*Carex praegracilis*	Clustered Field Sedge	FACW	Cunningham Marsh	6/26/1947
Cyperaceae	*Carex stipata*	Stalk-Grain Sedge	OBL	Sebastopol	6/6/1937

Family	Taxon Name	Common Name	Wet-land Status*	Locality	Collection Date
Cyperaceae	*Carex vesicaria*	Lesser Bladder Sedge	OBL	2 mi ne Graton; Sebastopol Quadrangle	5/17/1936
Cyperaceae	*Cyperus eragrostis*	Tall Flat Sedge	FACW	at Sebastopol Crossing La Laguna	8/17/1946
Cyperaceae	*Rhynchospora californica*	California Beak Sedge	OBL	Pitkin Marsh, 5 mi. N of Sebastopol	1/1/1936
Ericaceae	*Rhododendron occidentale*	Western Azalea	FAC	Sebastopol	7/12/1907
Fabaceae	*Hosackia gracilis*	Harlequin Deer-Vetch	FACW	Roadside between Sebastopol and Petaluma.	4/22/1947
Fabaceae	*Hosackia oblongifolia*	Streambank Deer-Vetch	OBL	Sebastopol	8/14/1908
Fabaceae	*Lupinus latifolius*	Broad-Leaf Lupine	FACW	2 mi. E mark west. T8n, r8w. N slope.	4/22/1934
Fabaceae	*Trifolium barbigerum*	Bearded Clover	FACW	Between Trenton and Mark West Mt. Oliver School; Santa Rosa	4/1/1929
Fabaceae	*Trifolium depauperatum*	Balloon Sack Clover	FAC	Near Windsor	4/18/1902
Fabaceae	*Trifolium grayi*	Gray's Clover	FACW	1 mile north of Windsor	5/10/1943
Iridaceae	*Sisyrinchium bellum*	California Blue-Eyed-Grass	FACW	Santa Rosa Creek	3/26/1902
Isoetaceae	*Isoetes howellii*	Howell's Quillwort	OBL	Sebastopol	6/6/1937
Juncaceae	*Juncus bufonius*	Toad Rush	FACW	E of Santa Rosa; Santa Rosa Creek	6/11/1902
Juncaceae	*Luzula comosa*	Pacific Wood-Rush	FAC	Santa Rosa Creek.	3/26/1902
Lamiaceae	*Pogogyne douglasii*	Douglas' Mesa-Mint	OBL	Lagoon at Sebastopol	7/1/1902

Family	Taxon Name	Common Name	Wetland Status*	Locality	Collection Date
Lamiaceae	*Prunella vulgaris*	Common Self-heal	FACU	1 mi NE Graton; Sebastopol Quadrangle	5/16/1936
Liliaceae	*Lilium pardalinum*	Leopard Lily	FACW	Sebastopol Road	6/21/1880
Limnanthaceae	*Limnanthes douglasii*	Douglas' Meadowfoam	OBL	6-7 mi w Santa Rosa (along road to Graton)	4/15/1938
Montiaceae	*Claytonia gypsophiloides*	Gypsum Springbeauty	FACU	Near Trenton	3/16/1902
Nymphaceae	*Nuphar polysepala*	Rocky Mountain Pond-lily	OBL	Sebastopol	9/27/1908
Phrymaceae	*Mimulus bicolor*	Yellow-and-White Monkey-Flower	FACU	Between Santa Rosa and Sebastopol.	6/8/1905
Phrymaceae	*Mimulus tricolor*	Tricolor Monkey-Flower	OBL	SW Santa Rosa (Stony Point Road, near The Lagunas)	5/19/1938
Poaceae	*Calamagrostis bolanderi*	Bolander's Reed Grass	FACW	Cunningham Marsh	6/26/1947
Poaceae	*Deschampsia danthonioides*	Annual Hair Grass	FACW	4 mi. W. Santa Rosa	5/18/1932
Poaceae	*Leersia oryzoides*	Rice Cut Grass	OBL	The Laguna just east of Sebastopol	8/17/1946
Poaceae	*Pleuropogon californicus*	California False Semaphore Grass	OBL	South of East Windsor	4/1/1949
Polygonaceae	*Persicaria hydropiperoides*	Swamp Smartweed	OBL	The Laguna north of Sebastopol	8/17/1946
Polygonaceae	*Persicaria lapathifolia*	Dock-Leaf Smartweed	FACW	The Laguna north of Sebastopol	8/17/1946
Polygonaceae	*Persicaria punctata*	Dotted Smartweed	OBL	The Laguna north of Sebastopol	8/17/1946
Potamogetonaceae	*Potamogeton illinoensis*	Illinois Pondweed	OBL	near Sebastopol	6/1/1899
Potamogetonaceae	*Potamogeton pusillus*	Small Pondweed	OBL	Near Sebastopol	6/1/1899

Family	Taxon Name	Common Name	Wetland Status*	Locality	Collection Date
Ranunculaceae	*Clematis ligusticifolia*	Deciduous Traveler's-Joy	FAC	Santa Rosa, Santa Rosa Creek	6/18/1937
Ranunculaceae	*Ranunculus californicus*	California Buttercup	FACU	Near Windsor	3/14/1902
Ranunculaceae	*Ranunculus lobbii*	Lobb's Water-Crowfoot	OBL	Sebastopol	n.d.
Ranunculaceae	*Ranunculus orthorhynchus*	Straight-Beak Buttercup	FACW	Near Windsor	3/14/1902
Ranunculaceae	*Ranunculus pusillus*	Low Spearwort	OBL	Sebastopol	n.d.
Ranunculaceae	*Thalictrum fendleri (var. polycarpum)*	Fendler's Meadow-Rue	FAC	Windsor	4/18/1902
Rosaceae	*Amelanchier utahensis*	Utah Service-Berry	FACU	Lagoon at Sebastopol	7/1/1902
Rosaceae	*Crataegus douglasii*	Black Hawthorn	FAC	Lagoon at Sebastopol	8/20/1902
Rosaceae	*Crataegus gaylussacia*	Suksdorf's Hawthorn	FAC	at Sebastopol	8/20/1902
Rosaceae	*Drymocallis cuneifolia*	Sticky cinquefoil	FAC	Santa Rosa Creek Canyon	5/5/1933
Rosaceae	*Holodiscus discolor*	Creambush	FACU	near Sebastopol	6/1/1998
Ruscaceae	*Maianthemum stellatum*	Starry False Solomon's-Seal	FACU	Santa Rosa Creek	4/15/1902
Salicaceae	*Salix laevigata*	Polished Willow	FACW	Santa Rosa Creek, 1 mile S of Santa Rosa	2/17/1931
Salicaceae	*Salix lasiandra*	Pacific Willow	FACW	Santa Rosa Creek E. of Santa Rosa	8/23/1902
Salicaceae	*Salix lasiolepis*	Arroyo Willow	FACW	E Santa Rosa; Santa Rosa Creek	3/18/1902
Sapiondaceae	*Acer negundo*	Boxelder	FACW	Santa Rosa Creek, Santa Rosa (bed of Santa Rosa Creek)	6/18/1937

Family	Taxon Name	Common Name	Wet-land Status*	Locality	Collection Date
Saxifragaceae	*Boykinia occidentalis*	Coastal Brook-foam	FAC	Mark West Creek	6/1/1886
Themidaceae	*Dichelostemma capitatum*	Bluedicks	FACU	Near Windsor	3/14/1902
Valerianaceae	*Plectritis congesta*	Short-Spur Seablush	FACU	Wright School on Sebastopol Road	5/13/1935
Violiaceae	*Viola adunca*	Hook-Spur Violet	FAC	Near Sebastopol	5/6/1899
Vitaceae	*Vitis californica*	California Grape	FACU	Mark West Creek	9/27/1907

*

Code	Indicator Status	Comment
OBL	Obligate Wetland	Almost always occur in wetlands
FACW	Facultative Wetland	Usually occur in wetlands, but may occur in non-wetlands
FAC	Facultative	Occur in wetlands and non-wetlands
FACU	Facultative Upland	Usually occur in non-wetlands, but may occur in wetlands

Reference: Lichvar, R.W., D.L. Banks, W.N. Kirchner, and N.C. Melvin. 2016. The National Wetland Plant List: 2016 wetland ratings. Phytoneuron 2016-30: 1-17. Published 28 April 2016. ISSN 2153 733X

Appendix C

Detailed Methods

This appendix provides detailed methodology used to calculate the landscape metrics and refine the maps shown in chapters 5 and 7.

PATCH SIZE AND NEAREST LARGE NEIGHBOR DISTANCE

The size distribution and nearest large neighbor distance of wetland and riparian habitat patches (features classified as "Valley Freshwater Marsh," "Wet Meadow," or "Forested Wetland or Riparian Forest/Scrub") was calculated from the historical, modern, and vision habitat layers.

In the GIS, discrete polygons within each habitat type were aggregated and considered part of a single patch if they were located within 60 m of one another (for marsh and wet meadow features) or 100 m of one another (for riparian features). Groups of polygons separated by less than this distance were identified and aggregated using ArcGIS's 'Aggregate Polygons' tool and assigned unique patch IDs. The 60 m threshold for grouping marsh and wet meadow polygons was taken from a rule set for defining resident intertidal rail patches developed by Collins and Grossinger (2004), which was based on the best available data on rail habitat affinities and dispersal distances. In the absence of more specific data, the rules developed for defining intertidal rail patches in the South Bay (primarily for Ridgway's Rail) were assumed to be broadly applicable to the Laguna's non-tidal freshwater wetlands/species. The 100 m threshold for grouping riparian polygons is based on the typical maximum gap crossing distance of dispersing songbirds (SFEI-ASC 2014).

The size of individual patches was determined with ArcGIS. In addition to determining the size of each patch, the number and distribution of "small," "medium," and "large" patches was also identified. For the purposes of this analysis, a patch was considered "large" if it had an area greater than 10 ha, "medium" if it had an area greater than 1 and less than 10 ha, and "small" if it had an area less than 1 ha.

Nearest large neighbor distance (NLND) for wetland and riparian patches was determined with ArcGIS's 'Generate Near Table' tool, which calculated the linear distance of each patch to the nearest "large" neighboring patch (>10 ha, see "Patch Size" above). Large patches themselves were assigned a NLND of 0 m.

RIPARIAN WIDTH

This analysis was used to visualize and quantify the length of riparian habitat based on its width. The width of riparian areas was determined by casting transects perpendicular to modified channel centerlines and then trimming the transects at the edges of riparian habitat polygons.

Riparian areas (features classified as "Forested Wetland or Riparian Forest/Scrub") were merged with adjacent "Perennial Freshwater Lake/Pond" or "Non-native Aquatic/Emergent Vegetation" polygons to create riparian habitat "zones". Next, centerlines were generated for each riparian habitat zone (from which to cast perpendicular transects that measure the zone's width at regular intervals). Channel polylines were used as the starting place for developing the riparian habitat centerlines, and were modified to adhere to the following rules:

- Riparian centerlines were not drawn for side channels within otherwise contiguous zones of riparian habitat.
- Riparian centerlines were not drawn for small lobes or splays.
- Riparian centerlines were straightened through sinuous areas.
- Riparian centerlines were smoothed with a maximum offset of 2 m and generalized by 0.1 to remove sharp angles and to prevent transects from being cast at incorrect angles.

The riparian centerlines were segmented at 50 m intervals and transects were cast perpendicularly from the centroid of each segment (as determined by the x, y coordinates of its endpoints) 2,000 m in each direction (a distance greater than the maximum width of the riparian habitat zone). Transects were then intersected with riparian zone polygons, thereby trimming the transects to the width of the adjacent woody riparian habitat zone. Transects were also manually trimmed in some areas, such as where multiple transects overlapped. Since riparian habitat "zones" included both woody riparian habitat as well as "Perennial Freshwater Lake/Pond" or "Non-native Aquatic/Emergent Vegetation" features, segments of the trimmed transects outside of "Forested Wetland or Riparian Forest/Scrub" polygons were erased to determine only the width of the woody riparian habitat. This process was automated with a custom ArcPy script.

CHANNEL-HABITAT ADJACENCY

The channel-habitat adjacency analysis was used to determine the relative composition of floodplain habitats adjacent to stream channels and open water areas. Channel polylines were attributed as either "mainstem" or "tributary" channels. A buffer of 5 m was applied to each side of the channel polylines to give the features an area. The resulting channel polygons were incorporated into the habitat layer with ArcGIS's Erase and Merge tools. These features were combined with features classified as "Perennial Freshwater Lake/Pond" in the habitat layer. The channel/open water polygons were then intersected with the habitat layer, resulting in a polyline that traces the locations where channel/open water touches other habitat types. The length of channel/open water adjacent to each habitat type was then calculated for all channels and for the mainstem channel alone.

TERRESTRIAL ZONES AROUND WETLANDS

This analysis characterized the composition of contemporary terrestrial habitats within an approximately 140 m buffer zone adjacent to wetland and aquatic habitats; the composition of the terrestrial buffer zone was not analyzed for the historical landscape due to lack of sufficient data.

Wetland and aquatic habitat types ("Farmed Wetland," "Perennial Freshwater Lake/Pond," "Non-native Aquatic/Emergent Vegetation," "Valley Freshwater Marsh," and "Wet Meadow") were combined and removed from the contemporary and vision habitat layers using the "Merge" and "Erase" tools in ArcGIS. Channel lines, buffered by 0.5 m to give them an area, were also included with the wetland and aquatic habitats. All habitat types not classified as wetland or aquatic were considered to be terrestrial habitat types. Although they include some wetland components, "Forested Wetland and Riparian Forest/Scrub" and "Oak Savanna or Woodland/Vernal Pool Complex/Valley Grassland" were considered to be terrestrial habitat types for the purposes of this analysis (which is consistent with the treatment of riparian areas in Semlitsch and Bodie [2003]).

A 142 m buffer was created around the wetland/aquatic habitat types, which is the minimum width identified by Semlitsch and Bodie (2001) as required to support semiaquatic species throughout their life history. The NCARI (SFEI-ASC 2014) and Sonoma Veg Map (Sonoma Veg Map 2017) datasets were used to classify land cover within the portions of the buffer zone that extended outside of the study area. Land cover classes in these areas were reclassified using the crosswalk in Table 5-1 on page 66; features with land cover classes not included in the crosswalk were assigned an "Unknown" classification (these represented a small fraction of the landscape). NCARI and Sonoma Veg Map classifications were also used to differentiate the habitat types in the "Oak Savanna or Woodland/Vernal Pool Complex/Valley Grassland" class. The buffer was then intersected with the modern and vision habitat mapping, and proportion of each habitat type within the buffer zone was calculated.

REFINEMENTS TO VISION MAP

A GIS analysis was performed to refine the opportunity areas identified by the TAC, and later refined by the MAC, within the Laguna to enable quantification of the increases in habitat area that restoration would achieve. The analysis was performed as follows:

PART 1: A refinement of initial, TAC-and MAC-identified areas using historical and modern land cover mapping and information about basic physical controls and suitability for different wetland types based on elevation, soil type and drainage class, and depth to groundwater was performed (Table 1). Example images of the part one process are below in Fig. C-1 and C-2.

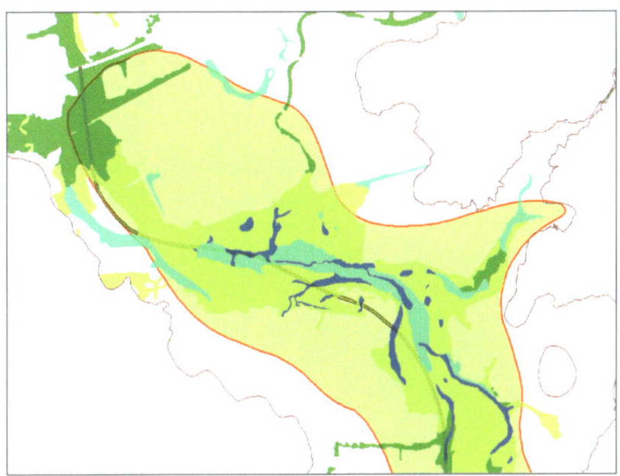

Figure C-1. Initial Wet meadow restoration opportunity area digitized from TAC meeting, outlined in orange. Existing wetland cover shown underneath restoration opportunity area for context.

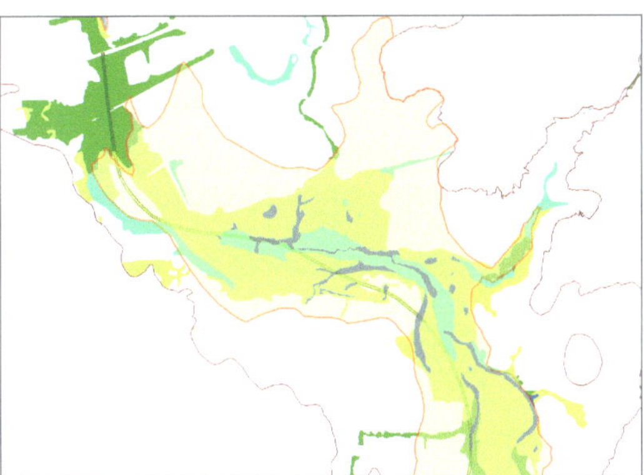

Figure C-2. Wet meadow restoration opportunity area refined in GIS, based on historical land cover and basic physical controls, outlined in orange.

Table 1. Restoration Opportunity Area Refinement Guidelines for Vision Map

Overall	Start with restoration opportunity areas identified by the TAC. Refine blobs by checking/reconciling with topography.
Open Water	Start with general extent of historical mapping of open water. Jonive: decision to include northern portion, which would convert existing wetlands to open water.
Freshwater Marsh	Start with general extent of historical mapping of freshwater marsh. Use areas of high groundwater and SSURGO soil type and drainage class as guides. Include areas that are poorly drained and have high groundwater (< 5ft).
Wet Meadow	Start with general extent of historical mapping of wet meadow. SSURGO drainage class/soils: Include contiguous areas with Drainage Class = "poorly drained". SSURGO soil type: acceptable soil types included (but not limited to) BcA, CfA, WmB, WoA, and CeA soil type categories
Willow Forested Wetland	Start with general extent of historical mapping of willow forested wetland. Check SSURGO soil types and drainage class. This type can exist in wet as well as loamy and better drained areas. Consult Q1 extent. Areas within Q1 more likely to be good willow forested wetland opportunities.
Vernal Pool	Start with general extent of historical mapping of vernal pool complex. SSURGO: Wha, WoA, HaB are soil type categories that support vernal pool.
Riparian Forest	Use line features to depict riparian restoration opportunities. Start with general historical mapping of riparian channels. For wide riparian buffer areas, consult Q1 extent. Areas within Q1 more likely to be good riparian forest opportunities.
Developed Areas	Remove developed areas depicted in the Sonoma Vegetation Map, such as mapped roads and buildings from the restoration opportunity areas.

PART 2: SFEI further refined polygons to exclude existing wetlands from opportunity areas. This is necessary to quantify the gain in wetland area that would be achieved through restoration. See Fig. C-3 below.

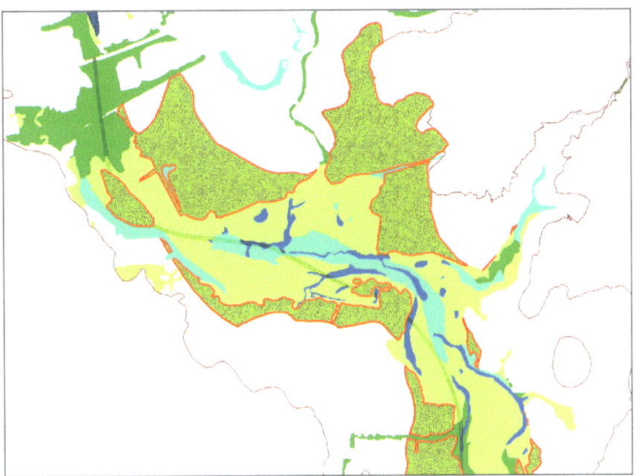

Figure C-3. Wet meadow restoration opportunity blobs (orange outline, stippled green), with existing wetlands clipped out to show opportunity area as expansion of existing wetland complex.

PART 3: Cleanup of gaps and overlaps, and removal of small 'orphans' that resulted from the erase procedure in Part 2 (Fig. C-4, C-5).

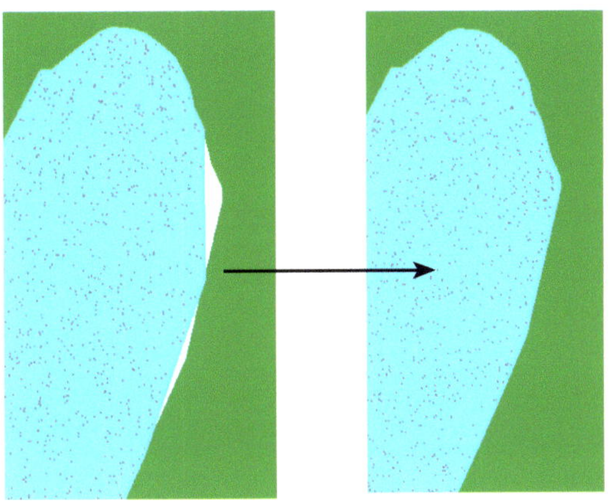

Figure C-4. Gap removal.

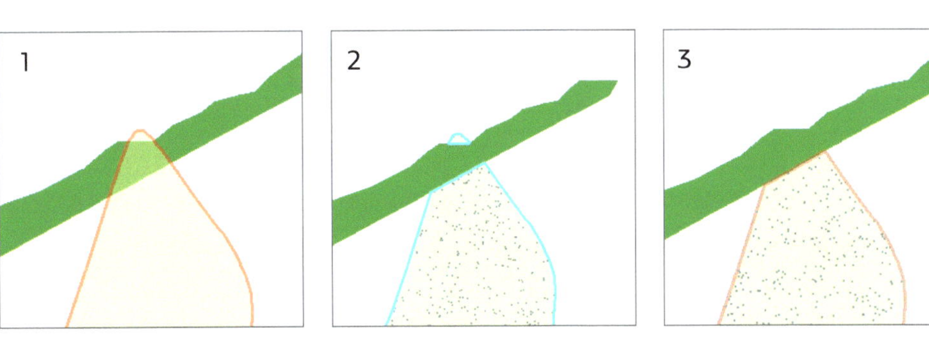

Figure C-5. Exclude existing wetlands from opportunity wetlands (E.g., Starting at (1), remove existing riparian forest, solid green, from light green opportunity area during Step 2 (2); and remove small, "orphan" artifacts to favor large, contiguous land cover types in Step 3 (3)).

www.ingramcontent.com/pod-product-compliance
Ingram Content Group UK Ltd.
Pitfield, Milton Keynes, MK11 3LW, UK
UKHW060106300726
14090UKWH00003B/388

* 9 7 8 1 9 5 0 3 1 3 0 8 2 *